THE ECOLOGICAL MONITORING PROGRAM (EMP) INDO-PACIFIC VERSION

STUDENT STUDY BOOK AND MANUAL

This manual has been published by Conservation Divers Ltd.

©2008-2019 Conservation Diver, all rights reserved

Written by Chad Scott

With Contributions by Rahul Mehrotra, Pau Urgell Plaza, Spencer Arnold, Florian Lang, Elouise Haskin and Nathan Cook.

All pictures, graphics, and text contained herein remain the sole copyright property of the author and may not be duplicated or distributed without explicit prior written permission from the author.

This book may not be reproduced, copied, or replicated without explicit written permission from the creator or Conservation Divers Ltd. Pt.

This manual only provides a background and a broad outline for conducting the Ecological Monitoring Program (EMP), this does not serve as an alternative to learning the course with an instructor. Anybody wishing to conduct the EMP should do so only under the close supervision of a trained professional to prevent harm to the environment or person. The writer of this guide and Conservation Divers Ltd. Prt. are not responsible for any injuries or problems sustained while participating in the Ecological Monitoring program. SCUBA diving is an inherently dangerous sport and is done at your own risk.

Please site this document as:

Scott CM (2019) **The Ecological Monitoring Program, Indo-Pacific Version.** Third Edition. *Conservation Diver Foundation,* Colorado, USA. 171pp.

Contents

Ecological Monitoring Program (EMP) Description 6
    Prerequisites 6
    Course Outline 6
  Certification Options 8
  Safety Notes For Scientific Diving 8

Chapter 1: Introduction to Coral Reef Ecology 11
  The Value of Coral Reefs 11
  Symbiosis and Interactions of reef organisms 12
  Coral Anatomy 13
  Requirements for a Healthy Reef 15
  Threats to Coral Reef Health 17

Chapter 2: Collecting Data Along a Transect Line 23
  Using Indicator Species 23
  Transect Lines 24
  Conducting the Invertebrate survey 30
  Conducting the Fish Survey 32
  Conducting the Substrate Survey 33
   Estimating Visibility using a secchi disk 34

Chapter 3: Invertebrate Indicator Species 37
  Introduction to Invertebrates 37
  Invertebrate Indicator Species 39

Chapter 4: Fish Indicator Species 57
  Fish Survey Indicator Species 57

Chapter 5: The Substrate Survey 73
  Substrate types and Codes – Non-Living 73
  Substrate Types and Codes – Living 76
  Hard Coral Growth Forms 79
  Coral health 86
  Review of the EMP Substrate Codes 90

Chapter 6 – EMP Dive Planning 93
  Required Equipment 93
  Pre-Dive Planning 94

The EMP Dive Procedure for 4 divers 95

Chapter 7: Coral Taxonomy 99

Introduction to coral families and Genera 99

How to identify coral genera 100

Identifying corals for the EMP survey 105

Common Coral Genera 106

Chapter 8: Surveying for Coral Diseases 119

Coral disease basics and terminology 120

Identifying problems with coral health 123

Coral Diseases of the Indo-Pacific 125

Chapter 9: Specialized reef survey Methods 135

Photographic Surveys 135

Quadrant Surveys 138

Benthic Surveys in Non-Reef Areas 139

Coral size-class, recruitment, and fragment survey 139

The Compromised Coral Health Survey (Disease Survey) 142

Mass Coral Bleaching Surveys 143

Advanced Fish Surveys 145

Chapter 10: Setting up a permanent transect line 150

Finding the transect starting points (point A) on existing lines 150

Setting up a new line 152

Mapping your new site 154

Appendix A 157

Fish and Invertebrate Example Slate Set-up 157

Appendix B 158

Substrates Example Slate Set-up 158

Appendix C 159

Substrate Cheat Sheet 159

Appendix D 160

List of Coral Genera (for Advanced EMP Surveyors) 160

Appendix E 161

Example Compromised Coral Health Survey Form 161

Appendix F 162

Coral Bleaaching Example Survey Slate 162

Appendix G 163

Dangerous Marine Animals of the Indo-Pacific 163

Glossary 168

List of Figures and Photo Credit 171

References and works cited 172

ACKNOWLEDGEMENTS

This manual would not be possible without the assistance of many people over the last 11 years. First and foremost, I must thank and acknowledge the Coastal Preservation and Development Foundation (CPAD Foundation) for developing the original Ecological Monitoring Program and training me and others on Koh Tao in this program. The work of Wimm Vadakan, Vichit Vadakan, Muay Anchalee, Watcharapol 'Kwan' Daengsubha, and Dr. Wayne Phillips of CPAD was the seed from which all of our current programs have grown from, and we are eternally indebted to their hard work, dedication, and enthusiasm.

Next, I would like to thank Devrim and Kaen Zahir for supporting and continuing this program when it seemed destined to be left behind. Their positive outlook and philosophies on life have brought me to where I am today. They are my mentors, my colleagues, my friends, and my family.

Lastly I would like to thank our most valuable contributors to the EMP data and program including Kirsty Magson, Amy Walker, Robbie Weterings, Caroline Leuba, Kai Christoph, Kelly Fisher, Matthew Harris, Jillian Dunic, Dagmar Albert, Heike Schwermer, Christoph Hoppe, Kaitlin Harris, Rahul Mehortra, Pau Urgell, Ploy Macintosh, Elouise Haskin, and Maria Fredrickson.

ECOLOGICAL MONITORING PROGRAM (EMP) DESCRIPTION

This course is designed to give experienced divers an opportunity to learn about coral reefs, perform a reef survey, and add to data used to monitor Koh Tao's coral reefs. By increasing awareness and involvement, we can decrease our impacts on coral reefs and provide solutions to protecting and restoring the reefs around the globe. Most of the successful projects involving restoring reefs around the world have been performed by small groups and communities, not governments or policy makers. We hope to continue this tradition and do what we can to protect the environment where we live.

After completing this course, students will know how to perform the reef research methods of the Ecological Monitoring Program and can continue to help with data collection. From the program, they will gain knowledge about coral reefs and their inhabitants. Students will learn to identify indicator fish and invertebrate species as well as substrate types, gain experience working with equipment underwater and also practice buoyancy and navigation skills. Students who complete the training can then volunteer to assist with the monthly EMP's and take their knowledge to other areas in the world.

PREREQUISITES

The following is the list of topics and requirements for the completion of the certification in the Ecological Monitoring Program. The prerequisites for beginning this recognition program are:

- Be an **Advanced level diver or have a minimum of 20 dives** or show proficiency in good buoyancy (your Advanced Course dives must have included Peak Performance Buoyancy, Deep Dive, and Navigation)
- Prior to diving the survey sites, all students must **successfully demonstrate proficiency in buoyancy skills**. Including but not limited to:
  - Basic Hovering
  - Diving with fins above head level
  - Non-disturbance of substrate (silt, etc)
  - Fin Pivot
  - Ability to be stationary without anchoring
- Review all Safety Procedures and understand the particular risks of the areas you will be diving
- Sign all health and liability waivers

COURSE OUTLINE

The following course outline is set for 4 days, which is believed to be the best amount of time to properly teach and give experience necessary for certification in the program. However, this schedule can be adapted to fit unique situations as long as all of the lectures and dives are completed in their entirety.

	Description	Topics/skills covered
Day 1		
Lecture	1. "Coral Reef Ecology" 2. "Collecting Data along a Transect Line: Invertebrates"	• Basic coral reef facts • Symbiosis and competition on the reef • Coral Anatomy • Requirements for reef health • Threats to coral reefs • Importance of monitoring programs • How to collect data using the roving diver and transect line techniques • Importance and identification of invertebrate indicator species
Dive	1. Conducting an invertebrate survey	1. Laying out a transect line 2. Identifying invertebrate indicator species 3. Recording data underwater 4. Reeling in a transect line
Day 2		
Lecture	1. "Collecting Data along a Transect Line: Fishes"	• Overview of phylogeny and evolution of coral reef animals • Identification of fish indicator species • Estimating fish size • Finding the starting point of a belt transect and navigate the line
Dive	Conducting a fish survey	1. Navigating a transect line 2. Identifying Fish Indicator Species 3. Recording data underwater 4. Measuring horizontal visibility using a secci disk
Day 3		
Lecture	1. "Collecting data along a transect line: Substrate" 2. Coral Growth Forms and Health	• How to conduct a point intercept transect survey • How to identify and code substrate types • substrate types as related to coral reef health and history • Coral growth form types • Identifying coral health
Dive	Collecting data along a transect line: Substrate	1. Identifying and coding substrate types 2. Identifying coral growth forms 3. Coding coral health
Day 4		

Dive	Conducting a full EMP Survey	• Navigating and laying out the line
		• Collecting data
		    ○ Invertebrates
		    ○ Fish
		    ○ Substrate
		• Measuring visibility using the secci disk

CERTIFICATION OPTIONS

The above description are the standards for the Ecological Monitoring Program under Conservation Diver. There are also two additional options. The first is for students who are not able to complete the full 4-day course, and is called The **Coral Reef Ecology and Monitoring** certification. Students receiving this certification must complete the Introduction to Marine Ecology Lecture plus one survey lecture (can be Invertebrates, Fish, or Substrate) plus complete at least 1 survey. If the student returns later, or visits another Conservation Diver program, they are welcome to complete the rest of the standards to achieve the Ecological Monitoring Program certification. The second option is for students who complete the EMP certification plus all of the research related certifications offered by Conservation Diver (see website for a full list and description) and is called the **Advanced Ecological Program** certification. This certification is not easy to achieve, and is only given to students who show exceptional underwater research skills.

SAFETY NOTES FOR SCIENTIFIC DIVING

Up to this point, you have been most likely been participating in only what is known as Recreational Diving. The EMP Program is an introduction into the world of Scientific Diving, designed for the general diving public. Similar programs are also used by reef researchers and students around the world, and so the course you are learning can allow you to help with professional efforts. There are, however, some things you must keep in mind when diving for scientific purposes rather than recreation.

Imagine if you told a group studying the rainforest that while in the field they can only stay for 1 hour, can't talk to each other while working, and if they drop something they most likely won't find it again. They would probably say it was impossible to work that way; welcome to scientific diving. Conducting research under the sea brings many challenges and requires good planning and judgment skills. In scientific diving, we are not focused on learning the SCUBA equipment or skills, but on using SCUBA as a tool to do something else. Many students find that this is much more enjoyable and rewarding than 'diving for fun', and will open your eyes to the world of coral reef animals and interactions that will interest you for life. Before beginning, please read the following safety tips and considerations.

- Follow all Safety precautions as a normal dive, paying extra attention to:
    - Maintaining Buoyancy at all times
    - Avoiding loose or dangling equipment which can catch on corals
    - Always dive horizontally with fins slightly above body to reduce stirring sediment
    - Check tank air pressure often
    - Use Surface Marking Buoys when appropriate
    - Watch for poisonous species, and know how to handle stings/bites (See Appendix G)
- Students observed to be contacting the reef, or not having inadequate buoyancy control will be asked to complete fun dives or practice skills before continuing further, those unable to improve will be excluded from receiving this certification.

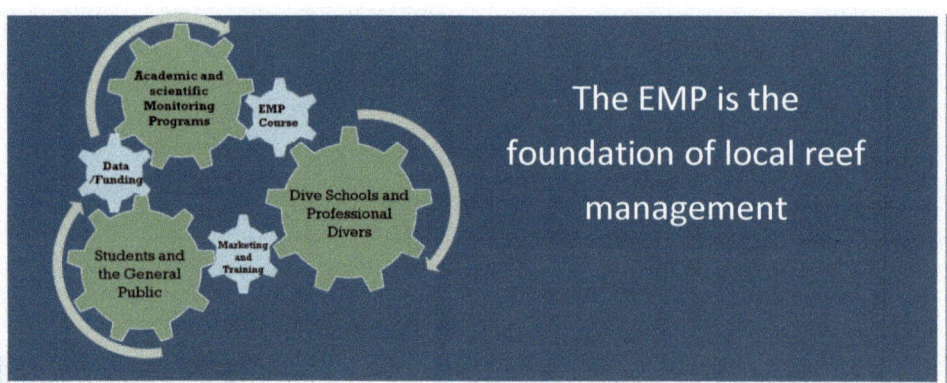

The EMP is the foundation of local reef management

Chapter 1: Introduction to Coral Reef Ecology

"Ecology is the study of organisms and their interactions with each other and their environment."

In order to monitor the reefs, you must have an understanding about how they function and what they require to be healthy.

- Basic Coral Reef Facts
- Symbiosis and competition on the reef
- Coral Anatomy
- Requirements for a healthy reef

Chapter 1: Introduction to Coral Reef Ecology

The Value of Coral Reefs

> **Corals form the foundation of the reef, provide habitat, and support the reef food chain.**

Coral Reefs are one of the earth's oldest, most diverse, and most productive **ecosystems**, hosting over 9 million species. Coral reefs account for less than 0.1% of the ocean floor, but are home to about 25% of all sea life, including sponges, worms, mollusks, crustaceans, reptiles, and fish. Coral reefs are aquatic ecosystems that are concentrated near coastal areas in shallow waters up to depths of approximately 50 meters. It is difficult to define all that constitutes what a coral reef is or what its boundaries are, but a complete definition of a coral reef is offered by Dr. Clive Wilkinson (1999):

> "A coral reef is a complex marine ecosystem of animals, plants and minerals in which most of the basal and vertical structures have been and are being constructed with calcium carbonate secreted by **hermatypic corals** and coralline algae, along with a variety of other carbonate- and silicate-secreting organisms."

Coral reefs have been called the rainforests of the sea due to their high amount of **biodiversity**, or different types of organisms present. Actually, coral reefs are more diverse than rainforests on a **phylum** taxonomic level, as rainforests are mostly composed of insects and flowering plants. Coral reefs are also one of the most productive environments in the world, this time tied with rainforests, sequestering about 4,000 grams of carbon per square meter per year. Most scientists would agree that life began in the sea, and it is in coral reefs where humans are able to see the effects of evolution in progress.

There are over **1,300** species of corals living around the world, each one unique

Coral reefs play a vital role in global ecosystem health and in many human economies. Coral reefs provide habitat and act as nurseries for fish and aquatic invertebrates, provide barriers from storms and waves to protect sea coasts, breakdown excess or **xenic nutrients** and compounds, and help to regulate atmospheric gases. The main structure of coral reefs is formed by hard corals, which are members of the Cnidarian phylum along with hydroids, jellyfish, and sea anemones.

Studies have found that about 2.6 billion people live within 100 miles (60 Km) of tropical coasts, which is projected to increase to 4.2 billion by 2050 (Sale et al 2014). Coral reefs have historically provided food, raw materials, medicines, and other services to substance economies around the world, and more recently have contributed greatly to national GDPs through reef tourism. In 1992, the world tourism market traded $1.9 Trillion USD, while the fishing industry traded about $27 Billion USD (Birkeland 1997). In other studies, the overall contribution of coral reefs to world economies is estimated at 375 Billion USD per year (West and Sale 2003, Costanza 1997). In South East Asia, many islands and villages are completely dependent on reef tourism to support their economy.

> **Coral Reefs are one of the earth's most valuable ecosystems – ecologically and economically**

Reef building organisms secret calcium carbonate ($CaCO_3$) skeletons which they create from dissolved carbon dioxide. In this way they serve as natural carbon sinks, playing a major role in atmospheric carbon sequestration and climate change stabilization. Without these climate buffers, predicted changes in the earth's temperatures would come about in a shorter time span, and possibly involve irreversible **positive feedback loops**. Indeed, in geologic history it was usually marine organisms which helped to lock up excess carbon dioxide from volcanoes and stabilize the planet's temperature after 'Hot Earth' events (basically the opposite of an ice age).

SYMBIOSIS AND INTERACTIONS OF REEF ORGANISMS

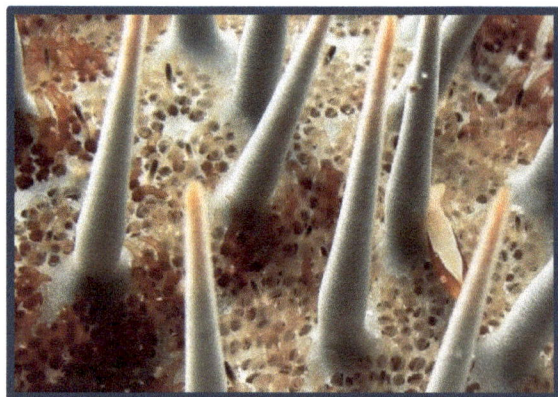

> **Ecology is the study of the inter-relationships between organisms and their environment**

One of the most interesting and valuable aspects of visiting the coral reefs is learning about their ecology. While **Biology** is the study of organisms, **Ecology** is the study of organisms, their interactions with other organisms, and with their environment. Much of what is known about ecology is derived from observation, and your job as an EMP student will be to observe and understand the complex webs of interactions taking place on the reef. It is more than a lifetime's work to understand them all, but once you learn the tools of observation you will see new and interesting relationships between animals every time you dive.

When talking about the interrelationships of animals, some of the first things that come to mind are predation and competition. These are two key mechanisms through which animals live, and it is important to understand the food chain of the reef to understand the balances that are in play. But, for the EMP we are more interested in understanding how different species of animals cooperate or interact in what are called **symbioses**.

Coral reefs exist based on a central theme of symbioses, based on co-evolution of organisms; most reef organisms are reliant upon another specific species of organism to survive. A reef is much like a city, where everybody has different jobs, and relies upon others to do their jobs (in a city if the subway driver doesn't show up - everybody is late). There are three main types of symbiosis that we will study:

MUTUALISM – A SMALL CLEANER WRASSE REMOVES PARASITES FROM THE SKIN OF A PUFFERFISH

1. **Mutualism** – A relationship in which both organisms benefit
2. **Commensalism** – A relationship in which one organism benefits, and the other is unaffected
3. **Parasitism** – A relationship is which one organism benefits and the other (the host) is at a loss

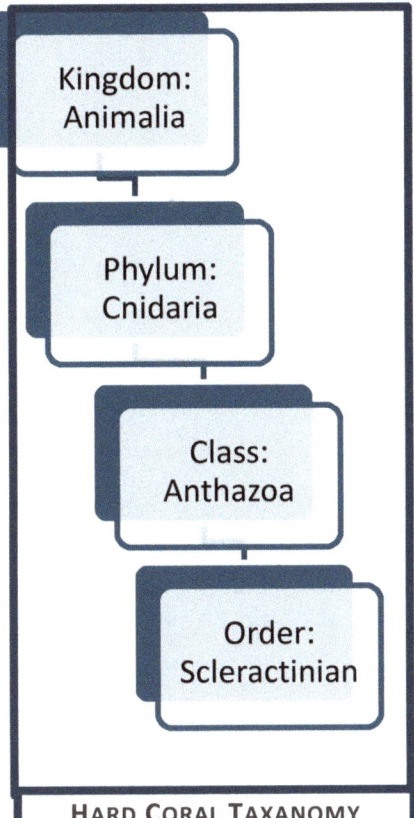

HARD CORAL TAXANOMY

Scleractinian corals are in the same kingdom as all other animals, including us. They are in the same phylum as Jellyfish and Anemones. And together with the soft corals make up the class Anthazoa.

CORAL ANATOMY

Corals are unique animals, much different from most other animals we see on land and in the sea in a variety of ways. In fact, for a long time people could not decide if corals should be classified as a rock, plant, or animal. Let's discover why all three could be considered correct when we talk about the classification of corals.

As mentioned before, a coral reef is a complex marine ecosystem based around the structure created by hard corals, calcareous algae, and the shells of many other organisms. A hard coral (called a **Scleractinian**) is then a smaller component of the reef; it is a colony of individual animals called polyps which together secrete a calcium carbonate skeleton. Coral skeletons grow in many shapes, called growth forms, that provide a variety of habitats for other animals such as fish. The polyps that make up a coral colony are small and resemble an overturned jellyfish. In fact, corals are in the same phylum as jellyfish, called the Cnidarian Phylum (along with anemones and hydroids). In evolutionary terms, this is a primitive phylum; some of the common characteristics of all the Cnidarians are:

- No Central Nervous System (brain)
- Lack of developed eyes
- Lack of digestive system (instead they have a digestive gut)
- 2 cell-layer thick skin (so thin it is transparent)
- Complex stinging cells (called Cnidi or **Nematocysts**)

Corals come in a wide variety of shapes and sizes, and so do their polyps. The basic body plan of the polyp can be thought of like a bag, with the top surrounded by long tentacles. These tentacles allow water to pass through, while the stinging cells stun and capture any prey which happen to pass by, primarily **zooplankton** such as krill and the larvae or eggs of other reef animals. The tentacles then bring the food to the mouth, and into the digestive gut. Unlike our digestive system, the coral has only one opening in the body, so the food must be eaten, digested, and then excreted before feeding can begin again (see figure 1 on next page). This is much less efficient than the digestive system of higher animals which allows for all three processes to occur at the same time, providing a continuous supply of energy.

Each coral polyp 'sits' inside of a small cup, called a **calice or corallite**, and is connected to all the other polyps in the colony through a layer of tissue called the **Coenosarc**. In this way, all the individuals work together for the growth and repair of the colony. They have a strong desire to do so because in fact every polyp in the colony has the exact same DNA, they are all clones.

A coral colony begins when one small polyp settles down onto a rock or existing structure, and then begins to grow and reproduce **asexually** until it runs out of room or encounters environmental constraints. As the coral polyps spread out, they also put down layers of

CLOSE-UP VIEW OF SOFT CORAL POLYPS

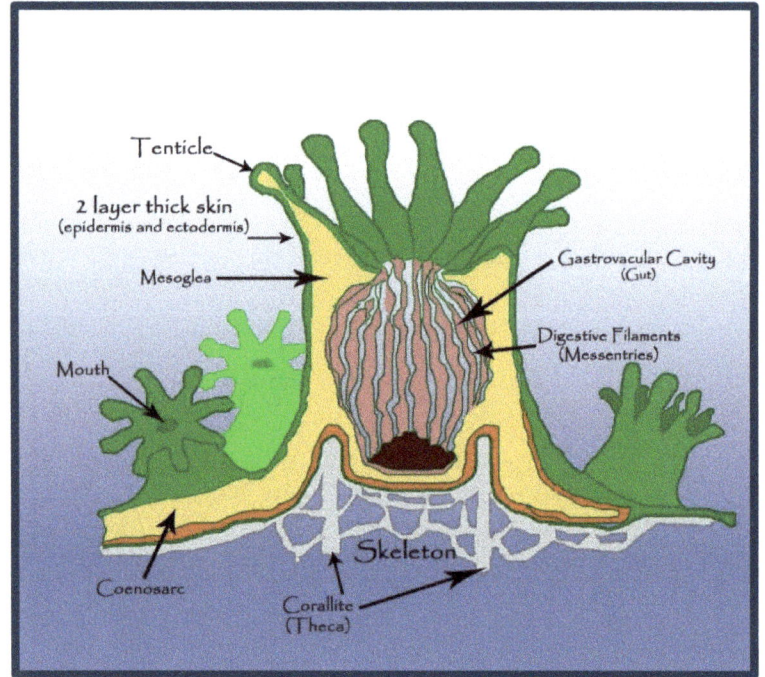

'glue' that become the skeleton, over time these layers can form the branching or brain like structures we see while diving. This process of building structure, called **bio-mineralization**, is very slow, and corals generally grow only 1-5 cm per year. Some very large corals can be thousands of years old.

The structure which corals put down is essentially limestone, and through the work of trillions of coral polyps over hundreds of millions of years, corals have created the largest structures built by any animals, including humans (the Great Barrier Reef is larger than the Great Wall of China.) The resulting structure is similar in strength to concrete. For humans, concrete production is the third largest contributor to green house gas emissions, due to the incredible amount of water and energy it takes to make it. How then, is it possible that these primitive coral polyps, with a very ineffective digestive process, create such structures? They have help.

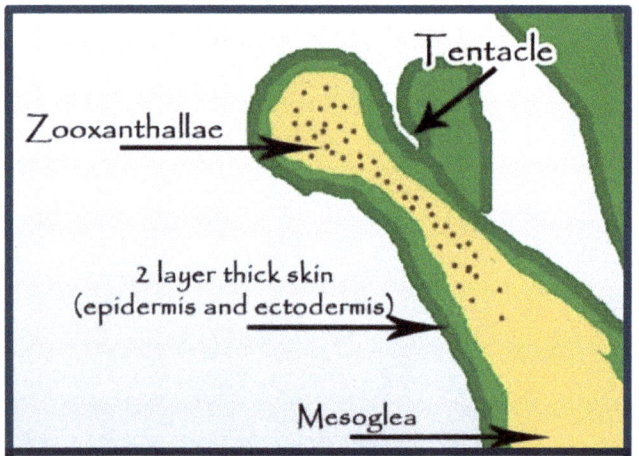

EACH TENTACLE ON THE POLYP IS COVERED IN HUNDREDS OF STINGING CELLS, CALLED NEMATOCYSTS, WHICH ARE USED TO CAPTURE PREY. ALL OF THE POLTYPS ARE CONENCTED TOGETHER THROUGH A LAYER OF TISSUE CALLED THE COENOSARC. THE ZOOXANTHALLAE LIVE THROUGHOUT THE MESOGLEA OF THE POLYPS AND TENTACLES (CLOSE-UP)

One of the most interesting symbioses in coral reefs, and the one that allows for all of the reef diversity and productivity, is that between the coral and a single celled **dinoflagellate algae** called **Zooxanthallae** (see close-up in figure 1). Recall that the coral tissue is so thin it is transparent, we should see right through the corals to the skeleton, and all the reefs should be a brilliant white color. When we see the corals as being green, red, brown, etc., the color we are seeing is actually these zooxanthallae, which fill all the space within the coral tissues. These tiny algal cells use the sun's energy to create sugars and carbohydrates, which they then share with the coral animal. In fact, for many corals 80-95% of the energy they use comes from their internal algal symbionts. When the coral is feeding it is really just getting nutrients to fertilize its algal 'crop'. In return, the algae receive a stable environment, protection from predators, and a steady supply of nutrients from the coral animal.

We can now see just how unique coral are in the world of animals. We can also see how the classification of corals can be quite tricky. Coral polyps are an animal; they move, feed, and reproduce sexually. Yet inside each animal is micro-algae (a plant) which use **photosynthesis** to create organic matter from light energy. And under it all is a mineral which is laid down so effectively it is found everywhere from the great ocean depths to the top of Mount Everest. When referring to corals, scientists often use the term '**Holobiont**' which in Greek means 'whole being' and refers to all three components of coral anatomy (animal, plant, and mineral). We can now investigate what is needed to maintain the health of the coral holobiont.

REQUIREMENTS FOR A HEALTHY REEF

Why, if corals are so productive and diverse, do they only cover less than 0.1% of the ocean floor? There are many factors which control, or limit, the areas where corals are able to grow and survive. In this section we will look at the main requirements for a healthy reef, which are:

- Light
- Temperature
- Water Quality
- Symbiosis
- And structure

> Light and temperature are the two most important factors affecting coral growth

LIGHT - As we saw in the last section, often up to 80-95% of the coral's metabolic energy supply comes from the work of photosynthetic zooxanthallae. Hard corals demand a high amount of metabolic energy to create skeleton in addition to all the normal energy demands such as tissue repair, immune system, mucous production, reproduction, etc. Because of this, hard corals are limited to the clear, shallow seas where there is a high abundance of light. Recall from your Deep Dive Course that light is lost in water due to reflection, refraction, absorption, and scattering. Therefore, most hard corals in the Indo-Pacific only grow in waters up to 30 meters deep; and sedimentation rates, or **phytoplankton** growth will decrease this depth dramatically. In low light conditions, algae within the corals are unable to produce enough energy to share with the coral animal, and both the algae and the coral are unable to survive. High intensity light, such as during the low tide, also stresses corals. The corals must use more metabolic energy to secrete a film that protects them against ultraviolet radiation (in their mucous layer) and also deal with the disposal of more free radicals formed during photosynthesis by their algae. The amount of light required (called photosynthetically available radiation or PAR) restricts the majority of hard coral growth to between 25° North and South Latitudes and to maximum depths of about 60 meters (recorded in the clear waters of Hawaii).

TEMPERATURE - The coral animal lacks thick skin, and thus their internal conditions are nearly the same as the external ones; they have no way to regulate their body temperature like we humans and most other higher animals do. Thus, corals are very sensitive to temperature changes, and are healthiest in a temperature range of 22° to 28° Celsius, where their enzymes and bodily reactions can take place with minimal waste. There are a few corals which can survive outside of this range, but it is rare for any hard corals to grow in waters that are within an 18º to 36ºC range. When temperatures are too low, the coral, much like us, cannot make

Mushroom Coral Bleaching on Koh Tao, 2014

metabolic reactions occur, and will gradually die. Higher temperatures can lead to more complicated problems which are a much larger threat to the coral's survival. When temperatures warm up, the environment inside the coral animal changes, it becomes more acidic and toxic to the zooxanthallae. Likewise, the zooxanthallae begin producing more waste and an excess of what are called free radicals (such as ozone and hydrogen peroxide) which attacks the coral animal. What is normally a mutualistic symbiosis breaks down, and becomes a competition for survival. The coral animal will then rid its body of the algae in a process known as coral bleaching (recall that without the zooxanthallae we see through the coral tissue to the white skeleton). Bleached corals are then living on only 5-15% of their normal energy budget, while they wait for better conditions to return or for a new type of zooxanthallae to move into them. If the coral does not have enough energy storage (fat and oils) to survive this period, it will die.

> **Coral Reefs require low levels of nutrients in the water or macro-algae will out compete the corals**

WATER QUALITY is the next requirement for coral health, and corals generally are able to grow in only clear water with low sediment and pollution concentrations. Corals require a **salinity** concentration in water between about 32 and 40 parts per million. Corals are sensitive to lower salinities and often do not survive near river deltas, and are affected by high run-off rates during storms. Sediment not only affects the amount of light available for symbiotic algae, but also creates stress to corals that must then use metabolic energy to produce mucus for protection. Pollution of waters can lead to diseases and create a wide range of problems for corals, most of which is not yet well understood. Coral larvae are extremely sensitive to even low amounts of pollution, and since coral larvae float at the surface, they are prone to being coated by petroleum products leaked from boats and terrestrial areas. Pollution not only degrades the living reef, but decreases reproduction and larval survival, preventing the reef from recovering.

Today, one of the biggest problems involving water quality is not chemical pollutants, but nutrient loading. Corals grow best in nutrient deficient areas, called **Oligotrophic** waters; this may seem confusing since we generally assume that a productive ecosystem must have a lot of nutrients (think of all the fertilizers that go on a crop of corn). But, recall that corals do not actually absorb nutrients from the water, they capture food, break it down, and give the nutrients from the food to their zooxanthallae (to fertilize their micro-algae crop). Where there are high levels of nutrients in the water, organisms other than corals can take advantage of them. The most destructive of which is **Macro-**

Algae, the plants of the reef (note that this is not the same as the micro-algae in the corals). Macro-Algae need many of the same things corals do (light and structure), but algae can grow much quicker than corals, over-growing them and smothering the reef. Corals generally do not grow in nutrient rich waters where quickly growing algae are at a competitive advantage. If water becomes nutrient rich (or low in algal grazers,) macro-algae can shade or smoother the slow growing corals (Castro 2007, Smith 2007).

SYMBIOSIS

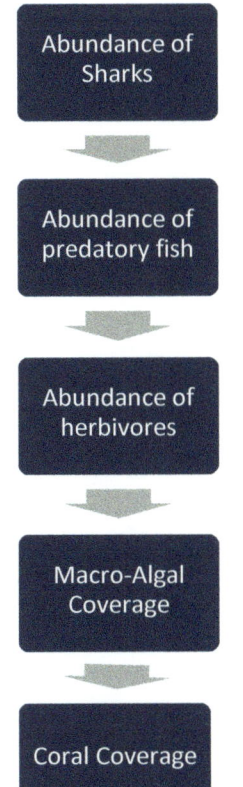

Corals also rely upon many fish and invertebrate grazer species to consume competitive algae. Even invertebrates and fish not living directly within the reefs help to balance the ecosystem. Oysters, clams, and sea cucumbers all eat algae and detritus from sandy beach areas before the nutrients in them reach the bays. Without these mudflat and sand dwelling organisms the corals would be overtaken by algae. Sea urchins, sea cucumbers, fish, and some snails will eat algae directly from the reef and open areas for coral growth. The elimination of any one species in a reef has far reaching effects throughout the coastal ecosystem. This is why symbiosis is the third requirement for reef health. We can look at this on many levels, which are often referred to as bottom-up and top-down controls. Bottom up means that the base of the food chain controls all that can survive at the top (i.e. the amount of fish controls the amount of sharks). Top-down means the reverse (i.e. the amount of sharks controls the amount of fish). The two controls work together to maintain the balance of the reef, for example: Sharks control the amount of predatory fish, which controls the number of herbivores fishes, which controls the amount of algae, which could otherwise out-compete the coral (thus sharks indirectly Influence the amount of coral).

The last requirement for reef growth, **STRUCTURE,** means that the corals need something solid to grow on. Coral larvae will not be able to settle or grow on sand and rubble, because when there are strong waves they will be turned over and killed. You will learn more about this aspect of reef health in Chapter 5.

THREATS TO CORAL REEF HEALTH

> **Today, coral reefs are one of the Earth's most threatened ecosystems**

Coral reefs are some of the most diverse, most productive, most valuable, and oldest continuous ecosystems on earth. Yet today, they are one of the most threatened. The biggest single threats to reefs are natural ones, such as tsunamis and cyclical ocean warming. One large natural event can almost completely decimate an entire reef, but these events are generally localized and infrequent; allowing the reef ample time to re-grow between events. The time scale for re-growth is on the order of 10-50 years providing that there are ample sources of diverse larvae in surrounding areas and that no new disturbances occur during this time. Anthropogenic effects tend to be less dramatic than natural ones, but are more frequent, widespread, and sustained (also called chronic), leaving no time for reef regeneration.

Human impacts to coral reefs that are considered the most destructive include introduction of organic and inorganic pollutants, increased sedimentation, and exploitation of reef resources. Specifically, these impacts stem from industrial,

> In 1998, *16%* of the world's hard corals died to due to bleaching and related threats

agricultural, and resort effluent or waste water, petroleum leakage from boats and cars, deforestation, and fertilizers. Also impacting corals are structural damages caused by boats and anchors, fishing nets, exotic fish hunters, divers, and sand or coral miners. Although these effects tend to be small and isolated, they add up to huge amounts of reef damage each year, and are chronic or reoccurring regularly.

The most threatening **anthropological** influence to coral reefs relates to climate change: alteration of the earth's atmospheric concentration of green house gases, increases in sea water temperatures, and changes in upwelling or ocean mixing. Most of the 36.8 Billion tons of CO2 emitted per year (2017) to the atmosphere does not stay there; it is absorbed by the oceans. Increased CO2 concentration alters sea water chemistry, and can reduce the calcification processes of corals and coralline algae by making the seas more acidic.

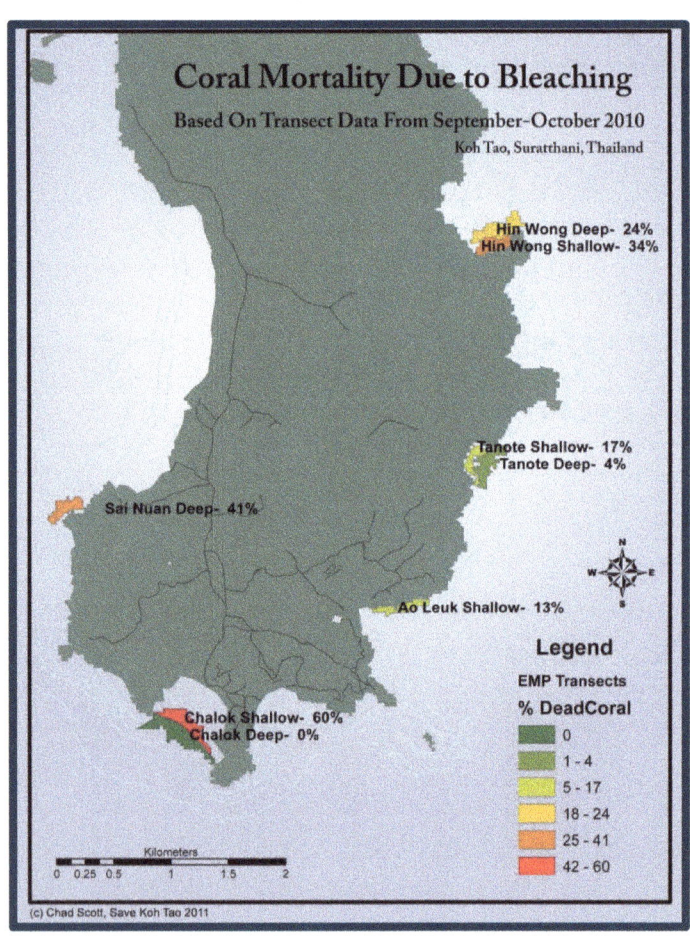

The 1997-98 El Niño-Southern Oscillation (ENSO) created the most widespread disturbance of coral reefs through increased sea temperatures in probably the past thousand years, 16% of the world's corals died in a single year. This was followed by more record breaking high years in 2002, 2004, 2006, 2010, and 2014-2016. This indicates that human influence on reefs is becoming more apparent and widespread, which is leading to lowered global reef diversity and limiting the ability for ecosystem succession by historical species. We are only beginning to realize the extent of our damage to reefs, seemingly mundane activities such as using sunscreen are now being shown to have far reaching effects on reefs. Climate change, in itself, will probably not completely destroy coral reefs if some biodiversity can be maintained. But, coupled with overexploitation and increased development, coral reef ecosystems could be degraded to the point where they are no longer productive or valuable to the ecosystems and economies that depend on them.

CASE STUDY 1: KOH TAO, THAILAND

Koh Tao is a small, 19km^2 island located in the Western Gulf of Thailand, which is surrounded by dense fringing reefs and several submerged pinnacles. Tourism, almost non-existent on the island 20 years ago, has boomed in recent years with between 300,000-400,000 tourists visiting the island each year by 2010, plus an unknown number of snorkeling or diving day trip boats from neighboring islands. The island accounts for the highest number of diving certifications issued for any location in Asia, and second in the world. Although the growth in tourism has brought economic wealth to the island, the terrestrial and marine ecosystems have been greatly stressed in the process. There is little historical data on the reefs of Koh Tao to evaluate past levels of abundance and diversity, however, reef decline had already been recognized in the region as early as the early 1990's. A 2006 study by Yeemin *et al.* found that there was a 17% decline in coral coverage on the island within a 5 year period, largely due to the 1998 mass coral bleaching event.

Coral Coverage On Koh Tao, 2008-2014 from EMP data

Location	n	2006	2007	2008	2009	2010	2011	2012	2013	2014	Mean
Hin Wong	60	52.5%	55.3%	62.3%	56.8%	41.3%	37.5%	33.2%	50.6%	71.8%	51.3%
Sairee	18	50.1%	*	62.5%	72.5%	*	*	40.6%	34.8%	43.1%	50.6%
Tanote	52	44.6%	*	22.3%	47.0%	42.3%	28.8%	23.8%	31.1%	33.6%	34.2%
Chalok	66	17.8%	16.3%	33.8%	22.3%	29.0%	18.0%	21.1%	26.4%	38.8%	24.8%
Aow Leuk	92	24.8%	*	21.6%	26.0%	28.2%	25.4%	21.8%	30.4%	30.4%	24.5%
Sai Nuan	39	17.1%	*	19.7%	28.6%	29.7%	25.9%	17.5%	16.5%	26.8%	22.7%
Totals	327	36.2%	27.8%	37.0%	32.0%	33.1%	27.1%	27.1%	32.6%	39.1%	

The table and graph above shows the change in coral coverage based on the average of all sites included in the EMP. Average coral coverage during 2006-2014 showed a 7.8% increase over the period, and ranged from 27% in 2011, to 39% in 2014. Of the EMP study sites, Hin Wong Bay has the highest coral mean coral coverage (51.3%) and increased 27% from 2006 to 2014. Had Sai Nuan has the lowest mean coral coverage (22.7%), but showed an increase of about 36% over the 9 year period. Three sites decreased in coral coverage during that time, losing between 16% to 33% hard coral cover (Mango Bay, Sairee, and Tanote Bay). Reef decline in 2008-2009 was in a large part due to the deforestation and construction of a large reservoir in the vicinity of Tanote Bay, which led to between 1-2 meters of silt and sediment inundating the reef of Tanote and increasing water turbidity as far south as Shark Island for 2-3 years. Decline in coral coverage from 2010-2011 can largely be attributed to the 2010 coral bleaching event, with more mortality at sites with anthropogenic stress such as sedimentation, coral predation, disease, or overgrowth by algae or tunicates.

On Koh Tao, many human activities have a direct and observable negative effect on reef health. Few waste water treatment systems exist or are functioning

> Over-fishing, development, and over-use are the main local stresses on many island's reefs around the globe

properly, most grey water flows directly into the environment, and sewage is held in septic tanks that are poorly built and rarely emptied. Showers, sinks, washing machines, and other grey water sources flow directly out of many resorts, homes, and restaurants into small surface ditches that lead to waterways or the sea. This causes an increase in nutrient and pollutant concentrations in the island bays, where it creates a competitive advantage for algae over corals; making rebound from bleaching events almost impossible in some areas. Development and deforestation releases tons of sediment into the coastal waters during rain events. This leads to reduced light availability for photosynthesis, causes further stress to corals for removal, and in some cases can bury reefs completely.

* * *

Giving you all of this information in Chapter 1 is not designed to make you feel depressed, but to help you understand the threats faced by the reefs and also tell you how we contribute to those effects. This is done so that as a diver you will be more conscious of the impact you leave behind, and also learn ways to make a positive impact to the areas you visit. **By taking the EMP Course you have already made a big step towards getting involved in the solutions to the problems our reefs face**. *Please be sure to also ask at your dive school what other research and restoration projects are available for you to learn about and participate in.*

A NOTE ABOUT DIVERS

Please remember not to touch corals or any other marine organisms. In studies conducted about the behaviors of divers that were observed without their knowledge, 222 divers were observed to contact the reef 129 times by accident, 38 times deliberately, and 55 times to anchor themselves. Over 75-95% of the contacts were from diver's fins, and 10% of the contacts were from other dive equipment. Reef contact was higher during night dives than in day dives, and there was an inverse relationship between diver experience level and the number of contacts. It was also observed that photographers contact the reef more frequently than any other diver type, about 4.5 times/10 minutes accidentally and 9.3 times/10 minutes deliberately (Walters 2007, Barker 2004). Also, re-suspension of sediment from diver's fins settles on the corals and causes stress to them. Koh Tao sees over 300,000 visitors per year, with more than half those visitors trying diving, and many taking at least one diving course (which involves 5 or more dives.) Remember that when doing the EMP you will be working close to the reef with extra equipment, watch your buoyancy and secure all of your SCUBA gear.

As divers, we are aliens into a foreign land; more people have been on the moon than to the great depths of the sea. In order to prevent unnecessary destruction to the reefs, most of which we do not immediately realize, it is vital that buoyancy skills be stressed at all times. Poor buoyancy skills can lead to bodily harm or destruction of corals. Just like humans, corals can get infections and viruses from wounds caused by diver contact. Swimming to close to the bottom can cause re-suspension of sediment that can settle on corals, and floating towards the surface leaves a diver prone to boat accidents. All students

An EMP Student shows perfect body positioning for conducting the EMP survey. Notice that it would be impossible for her to kick a coral or disturb the sands.

conducting the EMP must first show exemplarily buoyancy skills, and a buoyancy portion of the lecture is mandatory when teaching the EMP.

Often overlooked by novice divers is weight and body positioning while diving. Remember that the less weights you can use, the more comfortable you will be, and the longer your air will last. If your feet are usually below your head, then you have too many weights. The proper body positioning while conducting the EMP (or generally while diving) is to have your feet higher than your head. This ensures that you will not kick any corals, disrupt the sediment, or affect the study area. If you have not learned the frog kick, ask your instructor to show you, this technique allows you to control your position with more accuracy, as well as allows you to swim backwards.

Due to the amount of precision required for navigating transect lines it is recommended the EMP students practice navigation prior to learning the EMP. Almost more important that using a compass is to be able to use natural navigation and memory to locate buoys and move between research areas.

Chapter 1 Review

After completing the reading and discussion of the material covered in Chapter 1, you should understand and be able to answer the following questions. Please talk with your instructor about any questions you may have.

1. Why are coral reefs important (environmentally, economically, other...)?
2. What is Ecology? How is it different from Biology?
3. What are the three main types of symbiosis? Give an example of each.
4. What Phylum are corals in, and what are some of the common characteristics of all the members of this phylum?
5. Describe how the coral gets its energy, and some of the things that energy is used for.
6. What are the 5 main requirements for coral reef health? How does climate change affect these requirements?
7. Explain how algae can be both a vital helper, and a lethal killer for corals (hint: refer to Micro- vs. Macro-algae).
8. What are some things all divers can, and should, do to reduce their impact in the reefs?

Chapter 2: Collecting Data along a Transect Line

"Research and Monitoring are the heart of any reef management program, and require precision and practice."

Learning the methods and techniques for monitoring will help you further understand reef relationships, and how to view the reef in a more objective way.

- EMP Parameters
- Indicator Species
- Belt Transect Methods
- Recording Data

Chapter 2: Collecting Data Along a Transect Line

Introduction

Now that you understand a bit more about coral reef ecology, it is time to learn about how to apply that knowledge to study and monitor the reef. For monitoring programs like the EMP, data is collected on reef ecology using what are called **indicator species**. These are species which tell us something about the physical or biological factors at play on the reef. An example of a terrestrial indicator species would be the canary in the gold mine. Before sophisticated equipment for measuring air quality was available, miners would bring canaries into the mines with them as canaries are much more sensitive than humans to poor air quality. If the canary was alive, then the air is ok, if the canary dies then the air quality is bad, and the miners would get out quickly. People use some indicator species in their everyday lives, such as when you see flies or cockroaches and then assume the location you are is unclean.

> **Indicator Species tell us something about the physical or biological factors at play on the reef**

For the EMP we will use indicator species in similar ways. For example, if we visit an area that has a high abundance of macro-algae, then we can assume that there is an imbalance in the system, either too much nutrients or not enough herbivores. We could then look at the abundance of **herbivores** such as sea urchins. If urchin numbers are low then we could infer that the problem is over-collection, if their abundance is high then indeed nutrients could be the cause. In this way we can get a very good idea about the state of the reef without ever testing the water or sediments with high-tech instruments. By observing the symptoms and checking related factors we can trace back and identify the cause of many reef problems.

But how do we count reef animals? What areas do we count in, and for how long? And how can we ensure that we visit the same places every month? These are some of the questions that will be answered in this chapter.

Using Indicator Species

When we evaluate reef health through the use of indicator species there are two main things we look for; **Abundance** and **Biodiversity**. Abundance is the amount of an organism in a particular area. For instance, we may say that there are 65 sea urchins per every 100 square meters of reef. This does not mean much on its own, but we can compare this value to other areas, or track it in the same area over time. Through these methods we can get a lot of information on the related factors. Many sea urchins may be an indicator of too much nutrients (leading to an abundance of food) while too little may indicate habitat destruction or over-collection.

> **Reef Resilience is a factor of the health, abundance, and diversity of the corals and all the other reef organisms**

Biodiversity on the other hand refers to the variety of organisms. This can mean the number of different families, species, or even genetic individuals. In the case of the EMP we will use it to find out how many different species we have in each area, and assume that more species means a healthier reef. These two factors can be used together to infer coral reef resilience, which is the ability to withstand or recover from a disturbance. A reef which is abundant but not biodiverse (contains only 1 or 2 species of coral) is less resilient than a reef which is not abundant but is very diverse. This is because the more diverse reef can withstand a larger range of reef threats (sedimentation, pollution, bleaching, etc). Ideally, we want to see reefs that are both abundant and diverse, meaning that they have a high **resilience**.

Remember that there are over 9 million different species of organisms living in coral reefs, so we will not be assessing the abundance and diversity of all of them. Instead we will look only at our indicator species which have been chosen based on some of the factors below:

- **Feeding behavior**
- **Symbiosis**
- **Abundance and rarity**
- **Trophic level**
- **Susceptibility to threats/habitat destruction**
- **Popularity in fishing/collection**
- **Popularity amongst divers**

In Chapters 3 and 4 you will be learning to identify these indicator species, but first you must understand the techniques and methods used to collect the data.

TRANSECT LINES

Taking data on the reef is like taking a picture of the reef health. A picture can only show us a small portion of the reef in space and time, and if we try to compare random pictures from last year to this year it is not easy to know if the reef has changed or just the areas we photographed. But, if we took a picture from the exact same place every month then we could line them up like a movie and see the changes occur. Taking our data can be thought of in much of the same way. We will try to take data from the exact same place every month, so that our data can be correlated over time. To do this, we will take our data along a 100 meter tape measure (called a transect line) that is set up between two permanent marker buoys. Each month we will use the same buoys, and then have data which is always collected in the same place, in the same way.

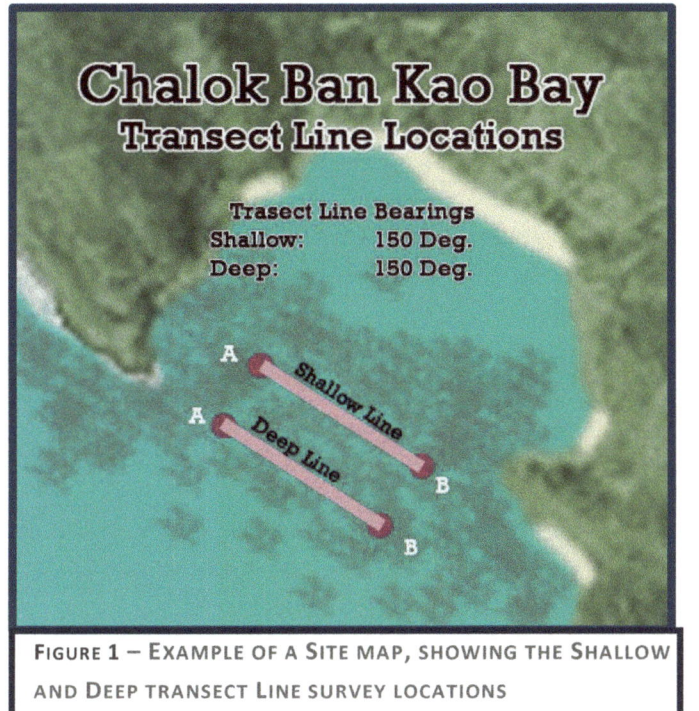

FIGURE 1 – EXAMPLE OF A SITE MAP, SHOWING THE SHALLOW AND DEEP TRANSECT LINE SURVEY LOCATIONS

LAYING OUT THE TRANSECT LINE

Each bay that is monitored contains two transect line locations, referred to as *Shallow* and *Deep*. This allows us to sample from two areas of the reefs and assess how changes occur over different depths or reef types.

At each line there is a starting marker and an ending marker (called 'A' and 'B', respectively) that the transect line will run between. The first task when conducting the EMP is to locate the Shallow A and Deep A points. For your course, your instructor will locate the buoys for you, but you should be aware of how natural features, triangulation, and search patterns are used so that you can relocate these points in the future.

After locating the starting point, the transect line is attached to the marker and each person in the buddy team takes on a different role. The first person becomes the navigator, and the second becomes the line manager.

The navigator has the most important job; for he or she needs to perfectly follow a pre-determined compass bearing to find the 'B' point (maps and bearings can be found in Chapter 10). If the B point is not located, then the data cannot be used. It is vital that the navigator maintains proper hand positioning with the compass, and use sighting of natural features to follow in a straight line. Even being off less than 5 degrees means that after 100 meters it may be impossible to find the ending point.

The navigator, being in front, must also occasionally check to make sure that the line manager is still ok, to do this, do not look over your shoulder, but instead look straight down through your legs. If you look over your shoulder you will move a meter or two off course. If there is a current you may also have to adjust your bearing or body position to compensate for it, this is where good training and practice become vital to completing the task.

The Line manager has only one job, to lay out the line behind the navigator. The line manager does not choose where the line goes. They must follow the navigator exactly - over corals, over rocks, etc. This is done to ensure that there is no researcher bias in the data, for instance, maybe this month a line manager lays out the line over the nice corals they like, and next month the line manager is scared of

damaging the corals and lays their line in the sand; in only one month it would appear all the corals on the reef disappeared.

When laying out the line, it is also important that you stay about 0.5 meters from the bottom. If you are too high then as the line falls it will not be straight, if to low you risk damaging the corals. It makes things much easier to do the survey if the metric side of the line is always face-up when it is laid out, but sometimes this is difficult to control. The line manager will most likely have to stop and tie the lines together in the middle if two 50 meter lines are used instead of a single 100 meter line. If the navigator is not paying attention at this time, the two may become separated.

RECORDING DATA ALONG THE LINE

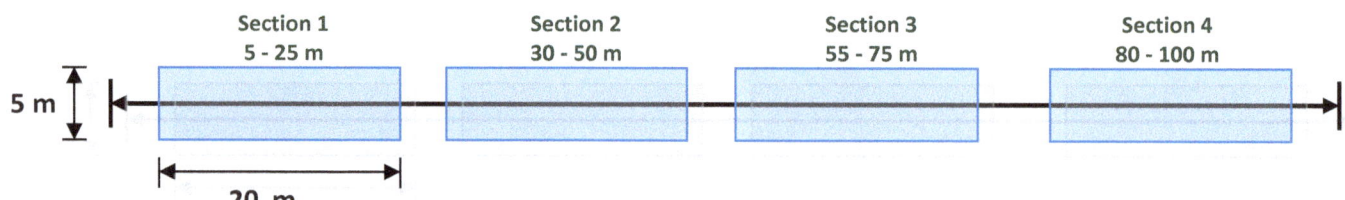

The drawing above shows how the transect line will be used to collect data for the EMP survey. The line is 100 meters long, and runs between the concrete markers known as 'A' and 'B'. The line is then divided into 4 segments. Data is collected in each of these segments, and then averaged together, so in reality you are doing 4 surveys of the reef, each separated by a 5 meter gap. These sections are each 5 meters wide (2.5 meters to either side of the line), and 20 meters long (totaling 100 square meters). This makes it very easy at the end to say there are X number of an animals per 100 square meters in this particular reef.

So, while conducting the EMP you will visualize the segments, and collect data only on animals found in the survey area. Any animals not inside the survey area should not be recorded, but can be added to the notes section at the bottom of your slate. Other things that can go in your notes section includes any interesting animals you see, any animals you don't know or want to ask your instructor about later, or any special occurrences you see on the dive, this may include:

- **Animals mating or spawning**
- **New animals not encountered on that reef before**
- **Thoughts or problems encountered on that dive that you want to talk to your instructor about**
- **Things to look for or remember when returning to the area (i.e. missing buoys, extensive damage to corals, debris or nets, etc.)**

The way that you record your data during the survey will be different for each type of survey completed. There are three survey types: Invertebrates, Fish and Substrate. Over the next sections you will learn the details of each survey, but first let's discuss some general considerations for recording data on your underwater slate.

THE EMP SLATE

To record your data along the transect line you will be using an underwater slate that has been set up especially for this survey. Be sure to check that your slate is up-to-date, and that no important parts have been rubbed out. At the top of the slate you will see the area for the **Metadata**; this means the data about your data, this includes:

- **The location (site name, be specific)**
- **Date**
- **Time of Day**
- **Transect (shallow or Deep)**
- **Your name**
- **The conditions at the site that day (Sunny, overcast, wavy, etc.)**
- **The visibility (very important but often forgotten!)**

Without the date or location your data is useless, and your effort and time will have been wasted, so be sure to fill out the top part of the slate before entering the water, as it is quite often forgotten at the end of the dive. Immediately after the dive, check your metadata in case there were any changes, and check over your slate to make sure it is legible and that you didn't have any questions about things underwater. It is much better to get everything checked while it is fresh in your mind then waiting until you return to the dive school.

You will notice that the slate is broken into a table that has the animals listed in the rows, and that there are 4 columns for recording data. Each of these columns refers to the 20 meter segments on the line, as discussed in the last section. Your data will be recorded using tally marks in the columns while you swim along the transect line. In some cases, certain indicator species may be extremely abundant, and making a mark for every ten is more manageable, above

EXAMPLE UNDERWATER SLATE SET-UP FOR PERFORMING THE EMP (NOTE, YOUR SLATE MAY DIFFER DEPENDING ON THE REGION WHERE THE SURVEY IS TAKING PLACE)

Ecological Monitoring Program Manual

all just be sure to have a system for marking them down which will allow you to accurately transcribe your values back on land.

It is important not to get to stressed or overwhelmed while conducting your first EMP survey, remember part of the idea here is to enjoy the work. Conducting these surveys while also concentrating on diving is difficult at first, especially for newer divers. However, most divers find that their buoyancy and self-awareness skills improve much faster performing these activities than it would if they were only fun diving. You will probably find that by being relaxed and aware of your body motions performing these surveys will become very rewarding and enjoyable.

Another thing to keep in mind, is that most likely we will not enter your first round of data collected into the database in order to preserve the integrity of the dataset. However, as you practice these techniques you will become more confident, and when you feel like you are doing it perfectly your data will be entered into the online database to contribute to our ongoing research. Those students who stay at the program for extended amounts of time will also receive instruction in data entry, however this is outside the scope of the normal EMP certification. If you are interested in knowing how the data is used, ask your instructor to direct you to some of the scientific papers that have been published based on the data collected through the EMP program.

SWIMMING ALONG THE TRANSECT LINE

Often the most difficult part for students on the first day is to keep track of which column to record the data in, and where they are on the line. Some tips that may help include:

- Keep your head down and your feet up, with the slate in front but not blocking your view, like in the photo at left; this will ensure you always have good orientation.

- Check your position on the line often

- Remember to stop recording data between the segments (the 5 meters gaps)

- Keep in mind that if you laid out the line you will begin the survey from the end (100 m) and work your way to the beginning (0 m) and must record data from columns 4 to 1

- Also remember that when using 2 x 50 meter lines, you will have to add 50 to whatever number you see on the tape measure during sections 3 & 4 [i.e. if you want to find 95 on the line - look for 45m on the tape (45 + 50 = 95)].

As you swim, you will look 2.5 meters on either side of the line to make the 5 meter wide survey area. To estimate 2.5 meters, you can visualize the average person laid out straight with their fins and arms extended. Your instructor will show you what 2.5 meters look like underwater before beginning the survey. You will record data for 20 meters, then stop recording data, swim ahead 5 meters, then start again in the new column on your slate. It is good practice to wait for your buddy at the end of every survey segment as too not get to spread apart. As mentioned before, the way you swim through the survey area will change for each type of survey that you do (Fish, Invertebrates, or Substrate) and is outlined in the next sections.

SAFETY TIPS WHILE PERFORMING ANY SCIENTIFIC OR RESTORATION TASKS UNDERWATER

1. If performing surveys in shallow areas, be sure to stay low enough that you would not get hit by a boat propeller and always use a surface marker
2. If you get lost or disoriented, swim around for a minute or two to try to find the line again rather than immediately ascending
3. Always ascend using a surface marker buoy, or attempt to find a mooring buoy along which to ascend
4. Look and listen for boats when ascending
5. Always keep track of your air and never go below 50 bar
6. Watch your dive limits when performing deeper surveys
7. Always ascend as slowly as possible and perform a safety stop
8. Always watch for dangerous or poisonous animals
9. Always pay attention to dive and safety briefings and be sure to tell your instructor is you have any issues or are uncomfortable with the task you have been given
10. It is recommended to wear a wet suit or rash guard to protect against stings

Conducting the Invertebrate Survey

FIGURE 1 – HOW TO CONDUCT THE INVERTEBRATE SURVEY. IN THE FIGURE, THE TRANSECT LINE IS SHOWN IN WHITE, THE SURVEY AREA IS MARKED IN GREEN (5 X 20 METERS), AND THE PATH OF THE DIVER IS SHOWN BY THE BLUE ARROW.

Invertebrates are the animals on the reef that do not have a backbone. In the next chapter you will learn about all the invertebrate indicator species, for now know that most of them tend to be towards the bottom of the food chain (shrimp, crabs, etc.) but some can also be considered 'top-predators' (Crown of Thorns Starfish, Trumpet Triton Snail, etc.), and that herbivores invertebrates such as the Sea Urchins and Sea Cucumbers are important regulators of algae and nutrient cycling on the reef. Invertebrates tend to be slow moving, or **sessile** (non-moving), so you will have plenty of time to count and record them. Invertebrate identification is generally the easiest of the three surveys for beginning students, so most likely your instructor will have you complete this survey first.

Almost all of the invertebrate indicator species are **benthic**, meaning that they live on the bottom (are non-swimming). Many are also **cryptic**, meaning that they stay hidden during the day time in the holes and crevices of the reef structure. This means that you will have to take your time to look over the whole survey area – including on top, under, and inside holes in corals. The invertebrate survey is often the slowest of the three surveys, especially in areas with dense coral reef cover. Take your time during the survey, but also bear in mind that you will have to finish all four sections and reel in the line before you get low on air.

As you move down through the individual sections of the line, zigzag back and forth across the transect line, looking no more than 2.5 meters out on either side of the line (shown as the blue arrow in Figure 1 on the last page). Any animal with half or more of its body in that area should be counted. It is good to establish a scale for what 2.5 meters looks like relative to your body. Before beginning the EMP survey, place your fins at 0m on the line and extend your body forward, next reach out your arms and see where 2.5m on the line is in relation to your hands. This will help you to confirm whether Giant Clams or other animals laying right around 2.5 meters should be counted or not.

Some other tips to keep in mind during this survey include:

1. Look all around each coral or rock, checking as thoroughly as you can - WITHOUT disturbing the substrate or contacting the reef. Never move any natural objects to look for indicator species.
2. Stay within the 5 x 20 m survey zone.
3. Continue moving forward as you zigzag back and forth across the line.
4. Be sure not to count any animals twice.
5. Ask your instructor or draw a picture of any animals you are unsure about.
6. Quickly check over your slate at the end of each section to be sure nothing was forgotten.
7. Be sure to stop collecting data at the end of the section, and swim 5 meters before beginning again in a new column on your slate.
8. Wait for your buddy at the end of each section, and check each other's air levels before proceeding to the next section.
9. After you have finished all 4 sections and reached the end of the line, wait for your buddy and any other surveys and then reel the line in.

Always be sure to remember the safety and environmental considerations for diving during your EMP surveys. Although you may have increased distances from your buddies during the survey, you should always know where they are, keeping within visual contact. Be aware that time tends to go quickly when working underwater, and always be consistent in checking your tank pressure. You should also keep track of your tank pressure so that you can plan the tasks that you need to complete before the end of the dive. If you are starting to run low on air, do not try to finish the survey, instead let another buddy team know so that they can finish the data collection and reel in the line.

Conducting the Fish Survey

FIGURE 2 - CONDUCTING THE FISH SURVEY. IN THIS FIGURE, THE SURVEY AREA EXTENDS INTO THE WATER COLUMN 5 METERS FROM THE BOTTOM, CREATING A RESEARCH 'TUNNEL.' THE DIVER SWIMS SLOWLY AND STEADILY DOWN THE TRANSECT LINE.

For our purposes here, 'Fish' includes also all the **vertebrates** (animals with a back bone). So, for the fish survey you will also be looking out for Sea Turtles, Rays, Eels, etc. The vertebrates tend to be higher on the food chain, more variable day to day, and faster moving. This makes counting the fish much harder, as unlike the invertebrates, they move around and will swim away from the researcher. Your instructor will most likely have you do the fish survey after you have already had some practice with collecting data along a transect line through the Invertebrate survey.

Fish and other vertebrates can be either **demersal** (living near the bottom) or **pelagic** (living in the water column), so we will also have to look ahead, down below, and up above for the fish. This means that our survey area will be more like a tunnel than a belt, 5 meters wide by 20 meters long by 5 meters high (see figure 2 above). Furthermore, we have to add a time component to the fish survey, because counts will be very different depending on whether it takes a diver 5 minutes or 20 minutes to swim through the section. To do this, the diver should *go as slow as possible, without stopping*. Doing so will also help to slow down your breathing and movement so the fish will be less frightened by your presence. Your instructor will demonstrate the proper pace during your first Fish Survey Dive.

Some of the most important techniques and tips of advice for conducting the fish survey include:

- Swim slowly and steadily down the line, looking both far ahead and close around for different fish species.
- Keep track of where fish swim so that you do not count them twice.
- Estimate the number of fish in schools quickly but accurately.
- Do not count fish outside the survey area.
- Wait about 5 minutes after another diver has gone through the survey area before beginning to ensure that all the fish have not been scared off.

Conducting the Substrate Survey

FIGURE 3 — Conducting the Substrate Survey. In this figure, the point-intercept transect is shown. The white line represents the tranect, and the blue dots are the 'points' which will be sampled (every 50 cm on the line). Note the head down position of the diver.

Substrate refers to the bottom composition, or the sea floor. This includes both the non-living (**abiotic**) features and the living ones (**biotic**). The abiotic components of the reef area includes the rocks, sand, and dead coral. While the biotic components includes the living corals, the algae, and sponges. To conduct this survey, you will use the transect line as a guide to sample many points from the reef, these points can then be averaged to find out what the substrate consists of, and what the percentage of living coals is on the reef. Using the transect line removes your ability to choose which points to sample, and gives us the 'random' sampling method required for good data collection. This survey is the most difficult of the three, so your instructor will most likely teach it to you last.

The survey diver will swim down the line, stopping to record the substrate every 50 centimeters; this means you will sample 40 points per section (2 every meter for 20 meters). The point you are sampling is no bigger than a coin, so it is important that you perform this survey accurately, which requires practice with your instructor. The most important part of conducting the survey is ensuring that you look straight down on the line, 90 degrees to the tape. It is often helpful to think about having a weight on a string that hangs below the line and points to the area you are supposed to sample. If you view the point from an angle (shown by red arrows at left) than you will not record the point accurately.

To record the points, we do not write out the name of what is there, but instead use codes to make things easier and faster. You will learn the codes in Chapter 5 on conducting the Substrate survey. You will also learn how to assess and record the coral growth forms and health.

Some of the most important tips for conducting the substrate survey include:

- Orientate your body so you are facing into the current, with your head down and feet up.
- Swim slowly, checking often to be sure you are recording the correct location on the tape and slate.

- Do not hold the slate in front of your face, but instead keep one eye on the slate, and one on the reef to maintain your buoyancy.
- Remember to skip 5 meters ahead at the end of every 20 meter section.

Don't get frustrated your first time, it gets much easier with practice, and for most experienced EMP surveyors this is their preferred survey to perform.

ESTIMATING VISIBILITY USING A SECCHI DISK

When discussing the slate, you were asked to estimate the visibility. This is done through the use of what is called a Secchi disk. The Secchi disk allows us to accurately estimate the visibility when used horizontally, or what is known as turbidity when used vertically. This is essentially a measurement of how far light penetrates through the water, which is affected by many factors, including sedimentation, nutrient and phytoplankton concentration, and sun angle.

Estimating visibility using the Secchi disk is done by two divers, ideally at a depth of 5m. The first diver stays at 0 m on the transect line (point A) or alternatively holds the disk and the end of the measuring tape. The second diver swims away from

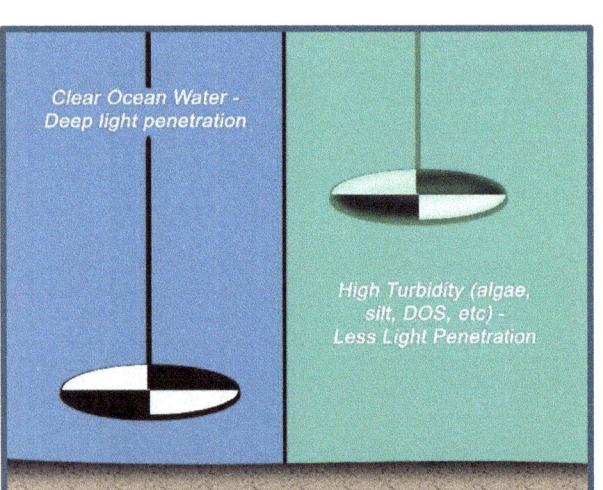

the first diver, keeping an eye on the disk. When the diver can no longer see the disk, they record the distance from it (this is the max distance). Next, the diver slowly swims back in until they can see the disk again (min distance). Next, they will average the two distance to get horizontal visibility.

On days when the sea bed cannot be seen from the boat, vertical visibility can also be measured by attaching the transect line to the Secchi disk and lowering it into the water. Again, the maximum (when the disk can no longer be seen) and minimum (when the disk comes back into view) distances are taken as an average and recorded.

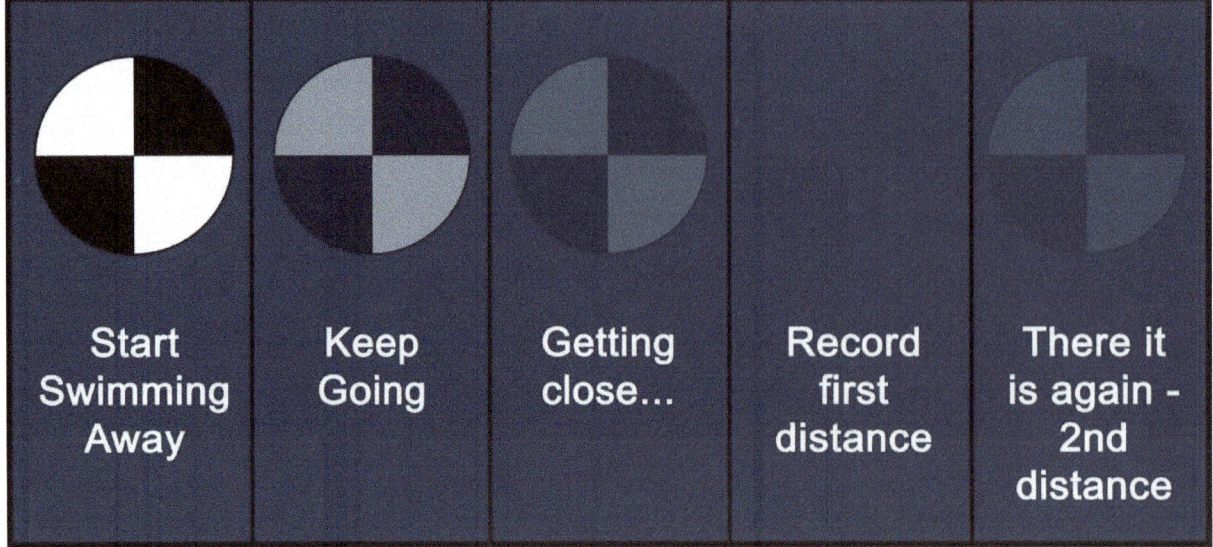

34 3rd Edition

You should now have an understanding of how transect lines are used to sample the reef during research and monitoring. You should also understand the differences between the three surveys for the EMP, and how each survey is conducted. In the next chapter you will learn about the animals and reef components that you will be assessing.

Chapter 2 Review

After completing the reading and discussion of the material covered in Chapter 2, you should understand and be able to answer the following questions. Please talk with your instructor about any questions you may have.

1. What are the parameters looked at in the Ecological Monitoring Program?
2. How are indicator species used? List at least three traits that an indicator species could display.
3. Draw, from memory, how the 4 segments are positioned on the transect line, label the dimensions of the survey are an the transect line with a starting and ending point for each segment (5 m and 25 m, 30 m and. . . .)
4. What is the main objective and rule for the diver navigating the transect line? For the diver managing the line?
5. How high above the substrate should the line manager swim when laying out the line?
6. Explain how to swim through the survey area when conducting the following surveys:
    a. Invertebrate
    b. Fish
    c. Substrate
7. What are 3 important safety consideration to remember while conducting the EMP?
8. Read Appendix G on *Dangerous Marine Animals of Koh Tao* and be sure you know how to identify each one underwater.

Chapter 3: Invertebrate Indicator Species

"Invertebrate phyla are the most diverse and abundant on Earth, they are the ancestors of higher life, and the first to adapt many of the body parts we take for granted."

- Evolution of Invertebrates
- The phylogenic tree
- Invertebrates Indicator species
  - Identification
  - Characteristics
  - importance

Chapter 3: Invertebrate Indicator Species

Introduction to Invertebrates

As mentioned in the last chapter, invertebrates are any animals without a backbone. If we view this group from an evolutionary standpoint, we would say they are more ancient than the vertebrates, which came later. This group includes some of the simplest animals on earth such as jellyfish, but also some much more physically and intellectually advanced animals such as the octopus. In fact, if we look at the evolutionary tree of animals, we find that many of these invertebrates are our very distant ancestors. Furthermore, many of the physical features we take for granted were developed in these invertebrate groups, including: the central nervous system, eyes, the digestive system, internal skeleton, and the circulatory system.

To learn about the invertebrates, we will look at the most biologically simple groups, and advance to the more complex. This will roughly follow the evolutionary development of these animals. This does not imply that they came in successive order, as the evolutionary tree is very complex, nor that evolutionary traits only arose at one time with a particular group. Instead we will use the evolutionary tree and the taxonomic classifications as a guide to understanding and learning about the invertebrates.

The simplest marine animals

Life on earth is thought to have arisen about 3.8 billion years ago, only about 800 million years after the earth formed. Earth at this time was a very inhospitable place with constant volcanic eruptions, an atmosphere containing no oxygen but lots of toxic gases, no ozone layer to block out harmful UV radiation, and many other threats. The only place where life was possible was in the sea, and one of the first successful forms of life was bacteria. Bacteria are simple, they are only a single cell, and do not have a nucleus, but they are the most abundant and probably the most important group of organisms on the planet. Of the bacteria, the most important in life's history is **cyanobacteria**, for about 3.5 billion years ago they evolved the ability to produce organic matter (sugars and carbohydrates) by capturing the sun's energy, which is always in great supply.

Although photosynthesis is a key advancement for life, survival as a single celled organism is very unpredictable; one minute they may be at the sea surface getting too much light, and an hour later the current has brought them down to the deeps where they receive no light. Success for the cyanobacteria really came once they began organizing in colonies, creating structures called **stromatolites**. These are calcium carbonate structures, which look like rock pillars, and are formed by millions of bacterial cells working together to secrete a glue which holds them in place. This allowed them to be in the same conditions every day, and to specialize to those

LARGE BARREL SPONGES LIKE THIS ONE REPRESENT THE WORK OF MANY UNICELLULAR ORGANISMS ALL WORKING TOGETHER

conditions. After several hundred million years, they became so successful that they actually created the oxygen atmosphere and the ozone layer of the earth today, which initially killed off most of the life on the planet.

Having an oxygen rich atmosphere allowed for much larger and more energy demanding cells to evolve, called the **eukaryotes,** of the fungi, plants, and animals. However, despite their increased complexity, life was still confined to being single celled organisms. It took another approximately 1 billion years for multi-cellular life to get going, but with a few advancements life took off in what is known as the **Cambrian Explosion** (about 530 million years ago), when most of the phyla of animals on the earth today originated.

The link between **unicellular** and **multi-cellular** animals can be represented in the sponges, which are a collection of various species of unicellular organisms all working together to create a single colony. Although it is hard to think of them as truly a multi-cellular animal, they do have a high level of organization. The first 'real' multi-cellular animals are the **Cnidarians** (jellyfish, anemones, hydroids, and corals) which were discussed in Chapter 1. Recall that although they had tissue and nerves, they lacked a digestive system, a central nervous system, and have no specialized organs. The next group we will discuss, the flatworms, improved slightly on this body plan, and are the first invertebrate indicator species you will learn about for the EMP.

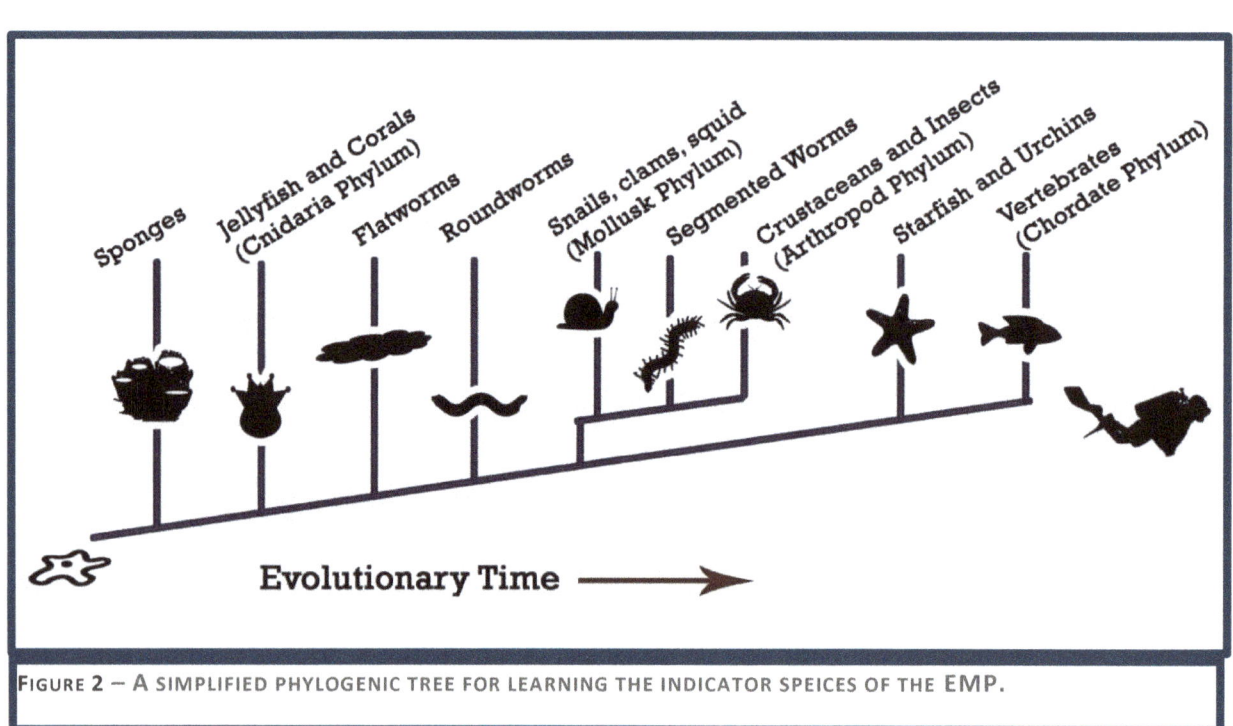

FIGURE 2 – A SIMPLIFIED PHYLOGENIC TREE FOR LEARNING THE INDICATOR SPEICES OF THE EMP.

Invertebrate Indicator Species

1. Flatworms (platyhelminthes Phylum, Turbellaria Class)

Flatworms are the next phylum after corals in our evolutionary guide to learning the invertebrates. Like the corals, they lack a digestive system, but they have advanced the ability to move around and hunt for food, instead of filter feeding. They also took the neural net of the cnidarians and concentrated it to the center of their body to form a more advanced control center (basic central nervous system). Furthermore, they concentrated nerves into balls to form primitive eye spots, which cannot see, but can detect changes in light.

Characteristics

Flatworms are indeed flat, and move around using an undulating motion of the body. It is difficult to differentiate the head from the tail end, and they do not have eyes or antennae (although some do have antennae-like folds at the front of the body referred to as 'ceros.') There are over 20,000 species, and most are very small and parasitic, but the ones found during the EMP are larger (about 2-10cm), and are free swimming carnivores.

Importance

Flatworms found on the reefs can be very sensitive to water quality changes and habitat destruction. Seeing flatworms is a sign of good reef health and diversity, having an absence of flatworms could indicate a larger problem on the reef.

Notes

Flatworms are mostly nocturnal, and are cryptic. They will be difficult to spot, and will generally be hidden during the day.

* * * *

Although the flatworms developed many advantages over the cnidarians, they still lack a digestive system. Like the cnidarians, the flatworm must eat, then digest, then excrete, before eating again. Meaning that unlike higher animals they do not get a constant supply of energy. A much more effective system of gaining energy

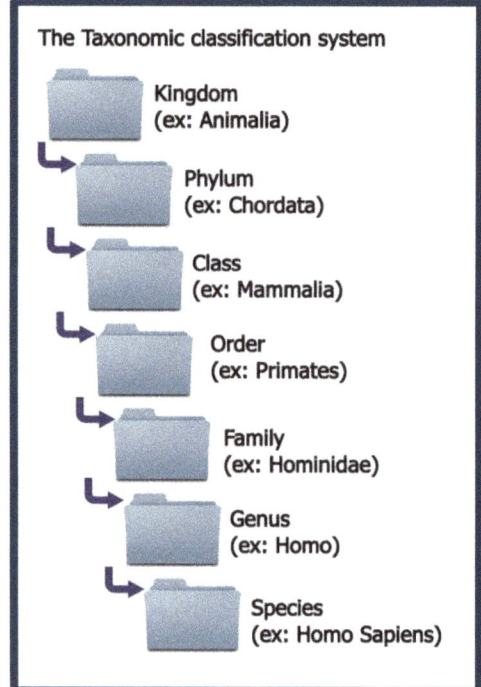

The Taxonomic classification system
- Kingdom (ex: Animalia)
- Phylum (ex: Chordata)
- Class (ex: Mammalia)
- Order (ex: Primates)
- Family (ex: Hominidae)
- Genus (ex: Homo)
- Species (ex: Homo Sapiens)

from food arose in the phyla of the roundworms (Nematodes). Along with the ribbon worms, these animals developed the complete digestive tract, allowing them to eat, digest, and excrete simultaneously, getting a constant supply of energy. Unlike the cnidarians and flatworms, round worms do have a circulatory system. Because they are generally hidden in the sand, we will not be monitoring nematode abundance during the EMP, however, we will look at some organisms which feed on them.

2. THE MOLLUSK PHYLUM

All of our next indicator species are part of the Mollusk phylum, which is the most diverse marine phylum on the planet, with up to 200,000 species, including snails, clams, and squid. Some of the common characteristics of this group include:

- Mantle (their special skin)
- Gills
- Foot (feet in the case of cephalopods)
- Radula (an efficient feeding apparatus, absent in bivalves)
- A shell (internal or external)
- And eyes

Because this group is so diverse, we are going to break it into classes to better understand and learn about this group as indicator species for the EMP:

- **Gastropods** (From Greek, *Gastros Podos* or "stomach footed") – these are the snails and slugs.

- **Bivalves** (meaning two valves, which are used to filter water) – these are the clams and mussels

- **Cephalopods** (from Greek *Cephal Podos* or "head footed") – these are the octopus, squid, and cuttlefish.

First, we will look at the gastropod class, which are the most diversified class of mollusks and exist on land, in fresh water, and in the sea. This class is made up of snails (those with a one-piece external shell) and slugs (those without an external shell). Although there are thousands of species of marine snails, we will only be assessing the abundance of 3 species for the EMP.

A. *DRUPELLA* SNAILS (CLASS: GASTROPODA, *FAMILY:* MURICIDAE)

CHARACTERISTICS

Drupella snails are a **carnivorous** gastropod which feed on coral tissue (called a **corallivore**). They are generally small (<5 cm), and are cryptic. They are long lived (up to 45 years) and generally occur in aggregations (groups).

IMPORTANCE

In small numbers, Drupella help to increase coral reef resilience by introducing small scale disturbances which improve diversity and open areas for coral settlement. But in high numbers, Drupella snails can consume vast amounts of coral, reducing diversity and abundance, and changing the population structure of the reef. It has also been shown that they are a vector for disease transmission, allowing coral disease to move from one colony to another. For the EMP we want to monitor Drupella numbers to watch for outbreaks or overpopulations.

NOTES

Because their shells are usually covered in calcareous algae, it is often very difficult to see the Drupella snails themselves. The best way to spot them is to actually look for their effects, the white coral of the reef which has recently been killed due to predation. As you conduct the survey, look for any clean, white coral skeleton, next look around the white skeleton to see if the predator is still around, very often you will find Drupella snails or the Crown of Thorns Starfish nearby.

Also, because it is too difficult to count Drupella, you will estimate their numbers during the EMP survey. At the end of every segment (before moving on to the next segment) estimate the number of Drupella that were in the survey area and write down a number between 0-4:

0 --> None
1 --> A Few (1-50)
2 --> Fair number (50-150)
3 --> Overabundance (>150)

In areas where there exists an over-population or outbreak of Drupella snails, collection programs may exist. By taking the Conservation Diver course 'Coral Predators: Population Monitoring and Management' you can become certified in these techniques and assist with those efforts.

B. Auger Snail (Class: Gastropod, Family: Terebridea)

Characteristics

The Auger Snail is a large carnivorous snail. They have a very effective toxin which they use to neutralize their prey, before consuming it with their powerful mouth, called a radula.

Importance

The Auger Snail is considered to be a top-predator, and thus important in top-down controls of the reef trophic structure.

Notes

The Auger Snail grows up to about 30 cm long, but lives primarily in the sandy areas on the reef edge. You will most likely not see the snail, but may see the track they leave as they move just under the sand, resembling a trail like in Bugs Bunny Cartoons.

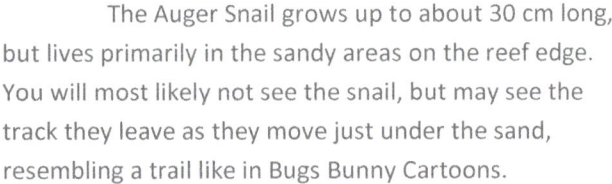

Some areas of the Indo-Pacific, host the Triton Trumpet Snail (*Charonia tritonis*.) These snails are a top-predator as well, and are one of the predators of the Crown of Thorns Starfish. In such areas, the Auger Snail (*Oxymeris maculate*) should be replaced with the Trumpet Triton on EMP surveys

C. Ramose Murex (Family: Muricidae)

The Ramose Murex, or Branched Murex is one of the largest and more ornate carnivorous gastropods in the Indo-Pacific. They tend to be found away from the reef, in sandy areas. You may find a higher abundance of them in areas with many sea urchins and sand dollars, as these are their primary prey sources.

Importance

As large carnivores, they are an important higher order predator that has few natural enemies.

They are an economically important species as their large shell is often appealing to the souvenir trade, and numbers are monitored to assess both biodiversity and the threats of over-collection. Due to their shape, they unfortunately are also prone to being entangled in lost fishing nets lying on the sea bed, one of the reasons why such nets should be removed by divers.

NOTES

The shells are generally covered in filamentous algae or other epibionts and fouling organisms (sponges, tunicates, etc.), and so the snails can often be difficult to differentiate from rubble. This becomes much easier with practice. During times of mating (September-October), large aggregations can be found in and around the reefs. In some cases, over 20 individuals can be found in a 100 m² area.

D. Nudibranchs and Other Sea Slugs (Class: Gastropod, subclass: Heterobranchia)

Characteristics

Nudibranchs and other sea slugs are a type of Gastropod very similar to snails, except that they do not have a shell (*Nudi* – Naked, *Branch*- lung, referring to the exposed gills otherwise under the shell in other gastropods). Other sea slugs may or may not have exposed gills, but have the 'slug' body shape and are not difficult to tell apart from other invertebrates on the reef. From an evolutionary standpoint, they most likely evolved from snails, but lost the shell overtime. A shell is great for protection from predators, but it is a big energy investment, difficult to carry around, and clumsy. Instead of the shell, Nudibranchs use other tools for protection; they feed on toxic organisms like sponges and hydroids and incorporate these defense mechanisms into their bodies for protection. You will distinguish the Nudibranchs by its snail shaped body, and exposed gills. They range from the almost microscopic, to over 10 cm in length.

the notes section when observed.

IMPORTANCE

Nudibranchs and sea slugs are incredibly diverse; they are a favorite amongst divers for their very striking colors and often ornamental bodies. The diversity of nudibranchs in an area is a good indicator of overall reef diversity. Furthermore, by monitoring the abundance and range of the animals it may be possible to better understand the effects of climate change on water temperatures and chemistry.

NOTES

You must look closely to find the Nudibranchs and other sea slugs, so take your time when doing the survey. Check around any of their typical food sources (i.e. sponges, hydroids, macroalgae, etc.) Remember, 'look but don't touch,' you are not permitted to move or overturn objects on the reef while searching for these or any other organisms. Also, be sure to look for the rose like eggs laid by the Nudibranchs, and record that in

If you are interested to learn more in-depth about Nudibranch, identify species, and help to monitor their populations, ask your instructor about the Conservation Diver **Nudibranch Ecology and ID** course.

E. Giant Clams (Class: Bivalves, Subfamily: Tridacninea)

Characteristics

The first Bivalve discussed here is the Giant Clam. This clam is a filter feeder, which also utilizes a symbiosis with zooxanthallae (the same as in corals) to build a very large and strong shell. You will always see the giant clam exposing its mantle (skin like tissue) up towards the sun, giving adequate light to their zooxanthallae crop. Because it is the zooxanthallae that give the giant clams their nice colors, no two giant clams are ever the same (like snowflakes).

Importance

The Giant Clam in Indo-Pacific reefs is considered to be a keystone species, or one of the most important species in maintaining ecosystem health. Some of their important traits include:

- **Filter feeding** – a single large Giant Clam can filter 1,000's of liters of water per day, removing nutrients and maintaining water quality

- **Contributing to reef structure** – The giant clam shell is very large (up to 1 m) and provides habitat and settlement points for corals and other sessile organisms. Often the shell provides a way for corals to begin growing in sandy areas.

- **Hosting Zooxanthallae**- By hosting zooxanthallae giant clams contribute to energy production on the reef, and also may help to 'seed' corals with zooxanthallae after a bleaching event

JUVENILE GIANT CLAMS IN ONE OF THE MANY NURSERIES MANAGED BY CONSERVATION DIVER TRAINING CENTERS

NOTES

There are many different types of clams and mussels growing on the reefs, be sure to only count the colorful giant clams and boring clams, both in the **Tridacna Family** and containing zooxanthallae (distinguished by their brightly colored mantle).

Although Giant Clams are protected by law throughout most of the Indo-Pacific, in many areas, their numbers are in decline due to over collection for food and shells. Due to declining populations around Thailand in particular, the Thai Navy, the Department of Marine and Coastal Resources and Royal Family have instituted a captive breeding and release program. The juvenile clams are raised in underwater cages before being big enough to move out onto the natural reef areas. Similar programs exist in other countries, and after completing the EMP you may be able to undergo additional training under Conservation Diver to participate in these programs.

F. Boring Clams (Subfamily: *Tridacninae*, Species: *T. Crocea*)

Characteristics

Boring clams are in the same family as the Giant Clams, but they bore into the rock and corals. So, while the Giant Clam is free living, these clams require substrate in which to grow. They contain the same photosynthetic zooxanthallae as the corals and giant clams, which allows them to be very productive. This also means that you will most likely find them on the top of coral heads where there is strong sunlight.

Importance

See Giant Clams.

Notes

In the Conservation Diver 'Giant Clam Nurseries and Population Studies' course you can learn much more about how to identify the various species within the family Tridacna. However, for the EMP we keep it simply as Giant clams (those free living with their shell exposed) and Boring (those without their shell exposed as they have bored down into the substrate).

G. Octopus, Squid, and Cuttlefish (Class: Cephalopoda)

Interestingly enough, cephalopods including octopus and squid are a part of the same phylum as clams and snails, even though they look very different. As mollusks they do have shells, but they evolved it to be internal; it is their beak. The beak of the Humboldt Squid is one of the hardest materials made by any animal, and has no problem biting through nearly any prey species. Cephalopods have also evolved incredible eye sight and intelligence, and also one of nature's most effective systems of camouflage. Not only are they able to change colors, but are able to change texture as well, imitating almost any reef surface. They also use their color changing ability for communication.

Because all of these animals play a similar predatory role on the reefs, they are lumped into one group together. If you do see any members of this class while performing the EMP also note the species.

CHARACTERISTICS

The Reef Octopus is very small, and inhabits the shallow reef areas. They are generally very well hidden and difficult to spot during the EMP. Squid and cuttlefish tend to stay in the deep sea during the day, and follow the zooplankton up at night to feed.

IMPORTANCE

These animals are quite rare, so observing them is a good sign of reef diversity. As a top predator, they require a large supply of food, and allow us to also gain information on the crustacean abundance around the island.

NOTES

Although rare in the reef areas surveyed for the EMP, if you see any cephalopods, including Octopus, Squid, and Cuttlefish please also take note on their behavior (hunting, hiding, mating, etc.).

* * *

The next group in the **phylogenic tree** [Figure 6] is the segmented worms. Although important, this is a very cryptic group, living under the substrate and inside the corals, and also being largely nocturnal. They are however very important in evolutionary terms, due to their contribution to the diversity of species.

Imagine that you needed to copy a page from a book; most likely you could do so with few mistakes, the copy may be different, but only slightly. This is what occurs during reproduction of an organism with a simple body plan, like round worms. Now imagine a segmented worm, their DNA must code for the development of each segment of the body, like a book with the same chapter repeated many times. In copying the whole book, many mistakes would inevitably occur, some which may be quite significant. In biology we call these random errors in DNA replication *mutations*.

Most mutations are not advantageous, and lead to reduced fitness of the organism. But occasionally a mutation comes along that helps the animal. Imagine our segmented worm, if one segment grew longer or more powerful legs it may hunt better. Over time advantageous traits become more prevalent in the gene pool, and new species can occur. If we look at the next phylum, the Arthropods, we can see just how advantageous segmented body parts can be for evolution.

THE TEREBELLID WORM IS A TYPE OF POLYCHAETE THAT LIVES IN CORAL REEFS

3. The Crustaceans (Phylum: Arthropoda)

THE MANTIS SHRIMP HAS THE MOST ADVANCED EYE SIGHT OF ANY LIVING ANIMAL, AND ONE OF THE MOST POWERFUL PUNCHES

Marine Crustaceans include the crabs, lobsters, shrimp, and barnacles. On land, this group includes all of the insects, one of the most successful and species rich groups on the planet. Like the segmented worms, they all have body segments and jointed limbs. Insects are so abundant that they make up about 80% of the animal species currently living on earth. They have jointed legs and external skeletons, which gives them great mechanical advantage, for example, the Mantis Shrimp uses its claws to break open clams, and can also break through aquarium glass.

Although crustaceans are extremely abundant and diverse in coral reefs, they are difficult to find and count. We can however indirectly assess their abundance by monitoring some of the many predators dependent upon crustaceans for most of their diet, such as the sting rays. One species that we can regularly locate and record on the EMP is the hermit crabs.

A. Hermit Crabs

Characteristics

Hermit crabs are specially adapted to fit inside the discarded shells of many marine gastropods. They use their specially adapted soft abdomens to grasp the inside of the shell, and thus utilize the shells for protection. The hard front legs of the hermit crab protrude from the shell, allowing it to walk around and feed.

Importance

Hermit crabs are important 'reef cleaners', much in the same way as ants in a rainforest. Hermit crabs feed on detritus or algae, acting as a major controlling factor on the abundance of macro-algae in the coral reef environment.

Notes

You will have to look closely at the gastropod shells that you see during the EMP survey to see if it is occupied by the snail or by a hermit crab. If you cannot tell, take your best guess – DO NOT pick up or overturn the shell to check. Some snail species are poisonous, and a sting can be fatal in only minutes. If you cannot tell, skip it or take your best guess.

* * *

Although an external skeleton is great for strength and protection, it does have some limitations. An Exoskeleton generally does not grow with the animal and it creates a barrier for gas exchange. For example, the lobster must dedicate a set of legs to pumping water through its thorax to its gills, and must shed its exoskeleton to

grow bigger, wasting resources and leaving it vulnerable to predation. An internal skeleton on the other hand, can grow with the animal and can also act as an additional organ in the animal (in us the internal skeleton is also responsible for creating our blood and immune cells). So, what is lost in mechanical advantage is more than made up for in other ways. The first example of an internal skeletal structure developed in our next phylum; the Echinoderms - or sea stars, urchins, and sea cumbers.

4. The Echinoderm phylum

The name Echinoderm means "Spiky Skin" and refers to the protective spikes of many species in this group. Because of their internal skeleton, all of these 'spikes' are covered in a tissue layer. Some of the common characteristics of this group include:

- Endoskeleton (Internal Skeleton)
- Water Vascular System (circulatory system)
- Tube feet (used for walking, detecting food, or feeding)

A. Crown of Thorns Starfish (Genus: *Acanthaster*)

Characteristics

The Crown of Thorns Starfish (COTs) is a large, corallivorous starfish with up to 19 legs. This very well protected sea star is covered in toxic spines, which only a few powerful predators such as the Triggerfish and Triton Trumpet are able to overcome.

Importance

As a large corallivore, in high abundances they can consume and destroy vast areas of coral, leading to decreased diversity and abundance of the reef. In low numbers (less than 2-3 per dive) they help to open clean coral skeleton for settlement of juvenile corals and other organisms, and strengthen the reef resilience. They can sense chemicals released by stressed corals, and thus remove less fit corals from the reef, much like a lion helps to keep zebra populations healthy by removing the weak or sick.

NOTES

Like Drupella snails, COTs can be found by looking for feeding scars, or freshly killed corals like in the picture at left.

COTs are on the rise in many areas due to overfishing and nutrient input. With less fish predation, and more nutrients in the water, more of the pelagic COTs larvae survive to adulthood, causing outbreak conditions in the worst cases. Throughout much of the Indo-Pacific, they have reached outbreak proportions, and are regularly poisoned or removed by teams from the diving community.

Do not try to touch or remove COTs while diving unless you are trained and equipped to do so, their spines are very sharp. Also, as they are covered in tissue, there is a very high chance of developing infections after being poked, which can have long lasting effects to your body (over 1 year in some cases). In certain instances, harming or disturbing the COTs will cause it to release its eggs or sperm in to the water (spawn) which can lead to increased population levels.

B. CUSHION STAR (GENUS: *CULCITA*)

CHARACTERISTICS

The Cushion Star has short arms which give it the appearance of a pentagonal pillow. Although it does not really look like other sea stars in its adult form, as juveniles they do resemble more typical sea stars. Its mouth and tube feet are all located on the underside of the body.

IMPORTANCE

The Cushion star feeds on detritus, small invertebrates, and also hard corals (Primarily fragments that are already under stress). It is monitored for its role in nutrient cycling on the reef, and also its potential to alter coral population dynamics.

NOTES

The Cushion Star is primarily nocturnal and will often be hiding under coral heads or rocks during the day time.

C. Long Spine Black Sea Urchin (Genus: *Diadema*)

Characteristics

Internally, urchins are quite similar to starfish, but instead their arms are fused into a round body. Long Spine Black Sea Urchins are very effective herbivores, able to eat much of the macro-algae on the reef which is unpalatable to fish. In areas such as Jamaica, where disease has reduced urchin numbers, algae becomes the dominate form of life in the area.

Importance

Sea Urchins are important in regulating macro-algae levels and contributing to the uptake of nutrients from the reef. They are a prey species for large fish and some other invertebrates (like the Horned Helmet). In high numbers they can also indicate an imbalance in either the nutrient levels (too much food availability) or decreased predator abundance (over fishing).

Notes

You will notice many different species of urchins on the reef but be sure to only count the ones with the long, thin spines and black body.

D. Marbled Sea Cucumber (Genus: *pearsonothuria*)

Sea Cucumbers (order: *Holothuriida*) are still Echinoderms, made up of many of the same parts as the urchins and sea stars, but they have an elongated body, and less distinctive spines.

Characteristics

The Marbled Sea Cucumber has modified tube feet surrounding its mouth, which are used as feeding appendages to scrape algae and organic matter from the rocks and corals. They have a black and white marbled body color, and have many small spines along the body.

Importance

Marbled sea cucumbers are important in regulating nutrient levels on the reef. They clean rocks and corals, removing microalgae and improving substrate availability for coral larvae and other organisms.

NOTES

In some areas of the world sea cucumbers, are considered a delicacy, or are used to supplement diets when food is scarce. This practice is incredibly destructive, as sea cucumbers are obvious, slow moving, and easy to collect. The populations of sea cucumbers of an entire reef can be decimated in a single day by an efficient team of collectors in dive gear.

E. BLACK SEA CUCUMBER (GENUS: *HOLOTHURIA*)

CHARACTERISTICS

The Black Sea Cucumber is smaller and less robust as the Marbled Sea Cucumber. They can be found mostly in sandy areas, or in the reef flat and back reef zones. They consume sand, and remove the organic matter or tiny invertebrates from it as it passes through their bodies.

IMPORTANCE

Black sea cucumbers are an important regulator of nutrients and organic matter flowing from land to the reefs. Their numbers reflect the available food supply, and thus are an indicator of water quality.

NOTES

To defend themselves, sea cucumbers can release their inner tissues, which are sticky and often toxic. This can protect them from predators, but comes at a large cost, as it may take over a month for the tissues to regrow.

F. PINKFISH SEA CUCUMBER (GENUS: *HOLOTHURIA*)

CHARACTERISTICS

The Pinkfish Sea Cucumber (also known as the edible sea cucumber) is very similar to the Black Sea Cucumber, but has a pink strip down the belly. They can be found mostly in sandy areas, or in the reef flat and back reef zones. They consume sand, and remove the organic matter from it as it passes through their bodies.

IMPORTANCE

Much like Black Sea Cucumbers, it is responsible for removing organic matter and biofilm from the sand to prevent algal overgrowth. Furthermore, as it feeds it aerates the sand, improving the conditions for burrowing invertebrates and allowing for aerobic conditions in which bacteria grow.

NOTES

The Pink Fish Sea Cucumber is differentiated from the Black Sea Cucumber as it is much more likely to be collected as food, especially in China or Indonesia.

G. ORANGE SPIKED SEA CUCUMBER (GENUS: *STICHOPUS*)

CHARACTERISTICS

The Orange Spiked Sea Cucumber is more robust, larger, and stronger looking than the Black Sea Cucumber. They inhabit sandy to rubble covered areas. Like the other sea cucumbers, they are detritivores who eat anf filter the sands.

IMPORTANCE

Much like the other sea cucumbers, they remove organic matter, biofilms, and small organisms from the sand as it is processed by their digestive system. They also help to aerate the upper layer of sand to prevent it from becoming anoxic.

NOTES

Sea cucumbers are important hosts for many parasitic or commensal fishes, shrimps, crabs, and worms. Sea cucumbers are also some of the fastest healers of the animal kingdom. This species can actually reproduce asexually when it is separated longitudinally due to its ability to heal and regenerate.

* * *

Through this chapter we have seen the development of life from its most primitive, through the development of most of the traits and features included in ourselves: digestive system, circulatory system, central nervous system, eyes, internal skeleton, etc.

The following list of species are not assessed in the EMP, however you may want to be familiar with them as they are commonly observed and sometimes confused with indicator species. If you find these species, you do not have to mark them down, however it never hurts to take good notes of all the interesting or unique animals that you see during your dives.

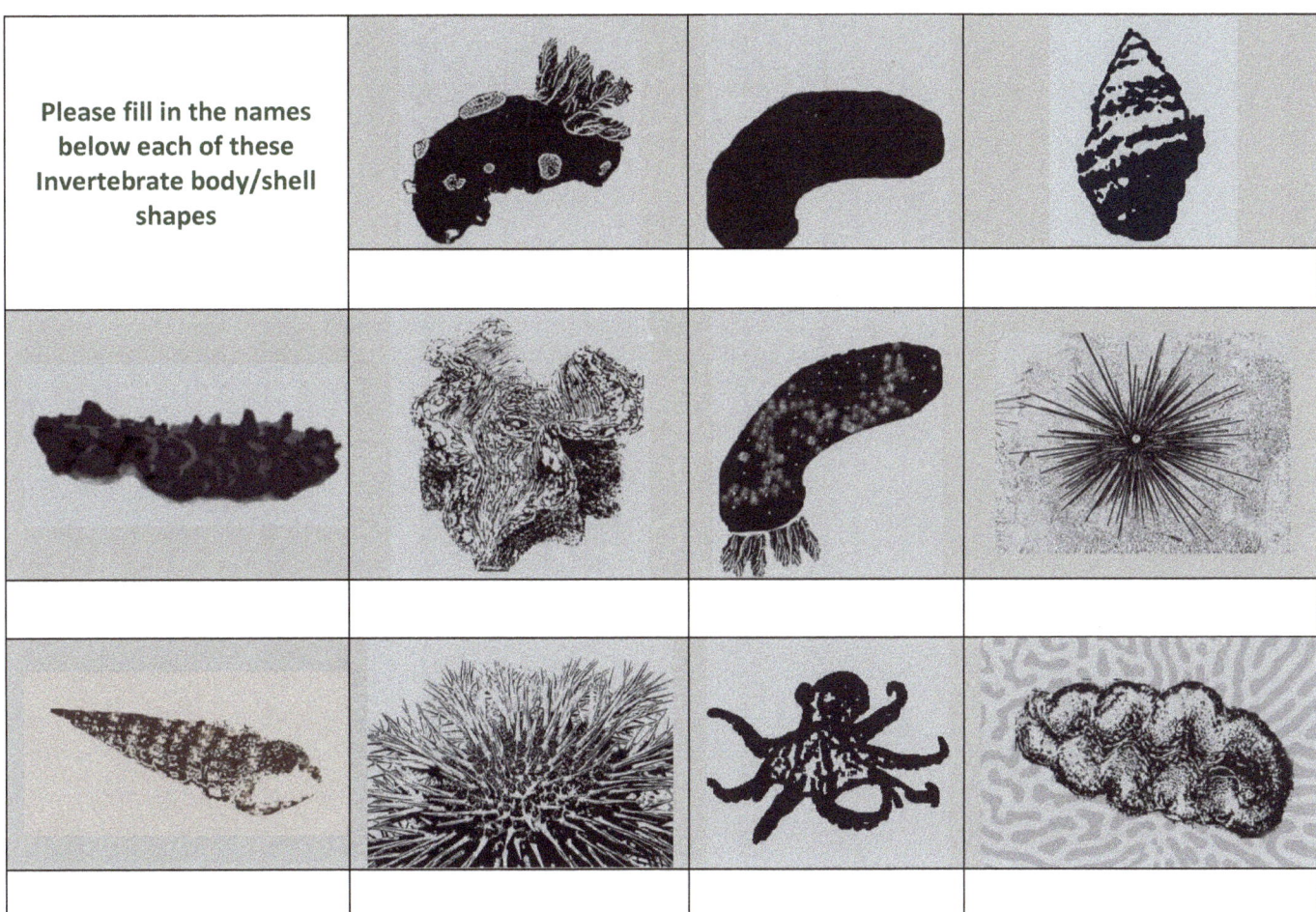

Chapter 3 Review

After completing the reading and discussion of the material covered in Chapter 2, you should understand and be able to answer the following questions. Please talk with your instructor about any questions you may have.

1. Which living animal represents the link between multi-cellular and single celled animals?
2. Explain how to record Drupella Snail numbers.
3. On a separate sheet of paper, please describe each of the following invertebrates. Also list at least two physical characteristics of each animal, and the ecological importance which makes this animals an indicator species. (try not to look back into the chapter if possible)

Flatworms	Drupella Snails	Ramose Murex
Triton Trumpet	Nudibranchs	Pink Fish Sea Cucumber
Giant Clams	Boring Clams	Octopus
Marbled Sea Cucumber	Crown of Thorns Starfish	Hermit Crab
Cushion Star	Black Sea Cucumber	
Long Spinned Black Sea Urchin	Orange Spiked Sea Cucumber	

Ecological Monitoring Program Manual

Chapter 4: Fish Indicator Species

"The most conspicuous animals on the reefs are the fish, like us, they are a part of the chordate phylum. Fish are our distant cousins, which could explain why divers have such an attraction to them"

- **Introduction to Vertebrates**
- **Fish Indicator species**
  - Identification
  - Characteristics
  - Importance

Chapter 4: Fish Indicator Species

Introduction to Marine Vertebrates

Through the invertebrate phyla we have seen life go from the most simple, to the more complex. The final phylum in our tree are called the Chordates. This group includes all of the fish, amphibians, reptiles, birds, and mammals. It is the phylum to which we belong. Some of the common characteristics of the chordates include; a hollow nerve channel running down the body, gills (many animals, like humans, only have these during fetal stages), a **notochord**, and a ventral heart. Most also have a backbone. With all of the components of life created through the invertebrate phyla (eyes, digestive system, central nervous system, etc.), plus with the more advanced and protected command system of the spinal cord, more advanced life took hold. Vertebrate groups include some of the most well adapted and intelligent animals, as we will discover in this chapter.

The first fishes evolved about 500 million years ago, the most primitive fishes that we find today are called jawless fishes, which include Hagfish and Lampreys. These fish lack the bone or cartilage that we associate with most vertebrates, and illustrate the link between our invertebrate and vertebrate groups. Slightly more advanced are the cartilaginous fishes, which is a very ancient group having a flexible skeleton made of cartilage instead of bone. This group includes families of fish such as the sharks and rays, known as Elasmobranchs. From there, species of bony fish, known as Osteichthyes, start to appear in the fossil records, becoming more diverse during the Jurassic period. Also during the Jurassic, marine reptiles such as sea snakes and sea turtles appeared in the geologic record. All of these groups include species that we will be looking for during the vertebrate surveys, which we refer to as the Fish Surveys.

Fish Survey Indicator Species

1. The Elasmobranchs

The first sharks evolved about 400 to as much as 450 million years ago, but many of these species were lost in the Permian extinction event (about 250 million years ago), which killed off as much ass 96% of all marine species. As life on the planet took hold again, rays and skates also evolved to join the Elasmobranch group.

Members of this group share the common characteristics of having 5 to 7 pairs of gills, placoid scales, and a jaw which is not fused to the skull.

A. Sharks

Sharks evolved as much as 420 million years ago, and have physically changed very little since then. Which is to say that in evolutionary terms they have not needed to change, for most of their history they have been optimally adapted to their environment. Today there are about 550 living species of sharks.

Characteristics

Sharks have a very streamlined, fast body (called **fusiform**), which is true of most predatory fish. They hunt using 6 senses; smell, sound, sight, touch (they sense vibrations in the water with their lateral line), **electroreception**, and lastly taste. Sharks, like humans, are late to reach sexual maturity and have very few, but well developed, offspring. In most cases, male sharks can be identified by claspers located in the pelvic region (like small 'arms' used for holding onto a female during mating).

Importance

Sharks are rightfully called the 'kings of the sea', because in most marine food chains they are the top predator, and thus are a top-down control on the entire ecosystem. Sharks maintain the ecosystem balance and they play a big role in nutrient cycling and export from reefs. The most common species of sharks you are likely to see are those associated with a coral reef environment such as Black Tip Reef sharks.

Notes

It is estimated that over 100 million sharks are killed by humans each year, and populations of most species of sharks are in sharp decline. Contrary to popular belief, sharks are not very dangerous to humans, especially in coral reef areas. The chances of seeing a shark during the EMP are not high, but you may be lucky enough to see them in the deeper dive sites you might visit between surveys.

B. Rays

Rays are closely related to sharks; they have **cartilagenic skeletons** and very advanced electrosensory abilities. Although there are occasionally Eagle Rays or Manta Rays in coral reefs in some regions, the primary type of rays found in coral reef areas is the Blue Spotted Ribbon Ray.

Characteristics

Sting Rays and Ribbon Rays feed primarily on benthic invertebrates such as crabs and shrimps. They are primarily nocturnal hunters, and will usually be found under coral heads or buried partially in the sand during the day. The Blue Spotted Ribbon Ray has a rounded body, and is easily recognized by its bright blue spots.

Importance

Rays are considered to be meso-predators (in the middle of the trophic structure), and so thus an important link between the top and bottom of the food chain. They are important in controlling populations of invertebrates, but are also a food source for higher order predators such as sharks. They are sensitive to changes in the environment, and also allow us to gain information on the abundance of crustaceans and bivalves on the reef.

Notes

Be sure to look well under the rock overhangs and coral heads for the rays during the fish survey of the EMP.

2. The Bony Fishes (Superclass: *Osteichthyes*, Class: *Actinopterygii*)

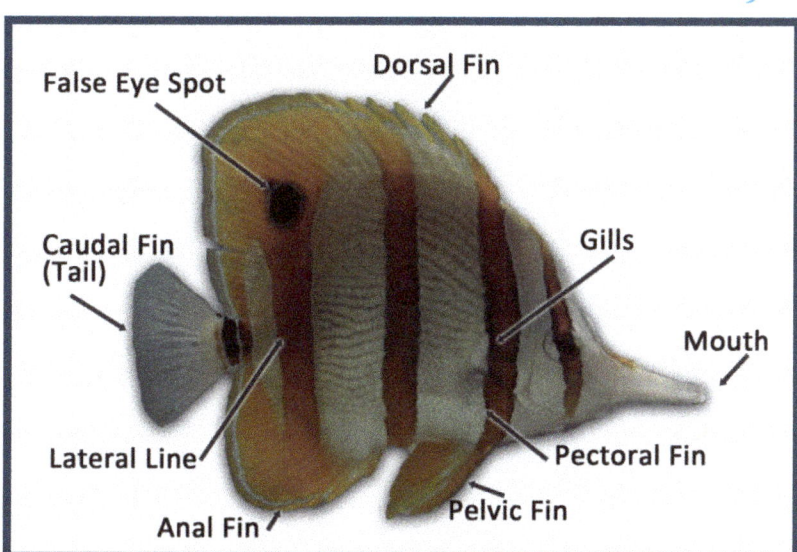

The next group of Chordates we will look at are the bony fishes, or more specifically the Ray-finned fishes. This is a very diverse group, consisting of over 435 families and about 28,000 species. Some of the common characteristics of this group include a primarily bone skeleton (rather than cartilage, they usually have an operculum covering the gills, and often have a swim bladder to maintain buoyancy.

Ecological Monitoring Program Manual

A. BUTTERFLYFISHES (FAMILY: CHAETODONTIDAE)

Our first group of familiar bony fishes found in the coral reef are the butterflyfishes. Most of the butterflyfish are very colorful and live in close association with corals. All of the butterflyfish have a similar body shape, which is plate-like (thin, tall, and round) and are of the order of perciform fish (meaning perch-like). They are the opposite shape of a shark's mouth, which illustrates how important a role predation has played in their evolution. Butterflyfish are the easiest fish to spot while conducting the EMP, and usually occur in pairs, but sometimes can be found in schools.

CHARACTERISTICS

Butterflyfishes are thin, tall, and plate-like (laterally compressed) to avoid predation. To further avoid predation, the tail of most butterfly fish looks just like the head, and often they have a line over their eye for disguise. Together these confuse predators, who don't know which direction to sneak up on the fish, or which way the fish is going to swim to get away.

IMPORTANCE

Butterflyfish only live in healthy reefs, graphs of coral abundance and butterfly fish abundance tend to be very closely correlated, and you can use one to estimate the other. Economically speaking, they are important for reef tourism, as they are a favorite of SCUBA divers and snorkelers. Although they are fished in some areas, they do not provide a good source of protein. In many areas they are a prized fish for the aquarium trade due to their bright colors, and are threatened by overcollection. Often times, cyanide or other poisons are used to stun and collect them, leading to the death of corals and many other reef animals.

Below are 5 examples of butterflyfish commonly used as indicator species for the EMP, which may need to be altered depending on the region you are monitoring in. Although their common names will vary from one region to another, we have provided the most popular common names here:

1. Weibels Butterfly Fish

Bright to dull orange with diagonal black stripes, it is one of the larger butterflyfish species. Note the black stripe over the eye and the similarity between the head and tail. This species is a true omnivore, feeding on a diverse diet of algae, zooplankton, coral polyps, small invertebrates, clams (including *Tridacna* species), and tubeworms. In some regions, it is also referred to as the Hong Kong Butterflyfish.

2. 8 Banded Butterflyfish

Yellowish white with black vertical bands, it is one of the smaller butterflyfish species. The juvenile and adult fish are nearly identical in coloration and shape. Juveniles will most often be found in table corals, and adults roam the reef to feed on coral polyps (this species is an obligate corallivore.)

3. Lined Butterfly Fish

White body with yellow tail, thin blue-black stripes running vertically down the body (looks like lined paper turned sideways). One of the largest butterflyfish species, they have a diverse diet of coral polyps, algae, and small invertebrates.

4. Copper Banded Butterfly fish

Another of the larger species included in our surveys, has a white-gold body with copper colored vertical stripes. Extended mouth facilitates a diverse diet of small invertebrates including worms, mollusks, and crustaceans. Note false eye (black spot) on tail for fooling predators. In some areas, referred to as the Beaked Coral Fish.

5. Long-Fin Banner Fish

White body with two vertical stripes and yellow tail. Easy identified by long white 'banner' and large pectoral fins. Feeds primarily on zooplankton, but also small reef invertebrates. In some areas they are called the Pennant Coralfish, and can sometimes be confused with the Moorish idol.

B. Groupers (Family: Serranidae)

Characteristics

Groupers are ambush predators; comparing them to the butterfly fish, they appear stronger, more streamlined, and faster. Groupers are generally **demersal** (live along the bottom) and can be found on top or under rocks and corals. They lie and wait for small fish or invertebrates to come close and then use their drawbridge like mouth to suck the prey in. Their lower jaw is spring loaded and opens and closes in a fraction of a second, but in doing so increases the volume of the mouth up to 500%, creating a current that few prey species can escape.

Importance

Groupers are predators, but are also a prey species for larger fish (mesopredators). They are a prized fish for human consumption, and are thus threatened by overfishing (especially the large species). Because of their behavior, they are one of the primary species caught through spearfishing.

NOTES

Although there are generally several species of grouper on any particular reef, you will not need to differentiate them by species. Instead, you will differentiate the groupers by size; larger or smaller than 30 centimeters. Your instructor will show you what 30 cm looks like underwater; also it is the same as the length of a standard EMP slate. The reason this is done, is like in many other fish species, they are protogynous, meaning that they change sex as they mature. All immature groupers are female, and only change to male in their adult life stage.

C. PARROT FISH (FAMILY: SCARIDAE)

CHARACTERISTICS

Parrot fish have a very fitting name for many reasons; they tend to be brightly colored, have fused teeth which form a parrot like beak, and swim with their pectoral fins in a bird-like fashion. Many species have the ability to secrete a mucus 'cocoon' while sleeping, which protects them from predation by electrosensory abled elasmobranch species.

IMPORTANCE

Parrot fish are one of the most abundant and effective fish grazers of micro- and macro-algae. They use their beak to scrape algae from rocks and dead coral which prevents it from overtaking the corals, opens up clean areas for settlement of coral larvae, and regulates nutrient levels in the reef. In some areas of the world, parrot fish can grow quite large and will actually use their beak to bite chunks off of the coral. This has given them the reputation as contributing greatly to bioerosion of the reef, however in those species (primarily Humphead Parrotfish) coral is not their primary food source.

NOTES

During the EMP you will differentiate the parrot fish in two groups, small (less than 20 cm) and large (greater than 20 cm). This is because, like the groupers, they all start of as females, and become males after developing to a certain age/size. Juvenile parrot fish tend be a duller green color, and are sometimes confused for the moon wrasse by beginning students. Often, in shallow reef areas, mixed schools of juvenile parrot fish and rabbit fish can be found. These are generally harems of females presided over by a single large male. During the EMP you should try to estimate the number of fishes in the school, but be sure to also differentiate between the parrot fish and rabbit fish.

Ecological Monitoring Program Manual

Double Barred Rabbitfish

Gold Saddle Rabbitfish

D. Rabbit fish (Family: Siganidae)

Characteristics

Rabbit fish are a common reef fish, and resemble the simple fish profile that everyone can recognize. They have a nose (rostrum) which resembles a rabbit's, and also play a very similar ecological role to rabbits; being prolific herbivores grazers. Their dorsal fin is composed of 13 spines, which are venomous. Being stung is painful, but not actually life threatening for humans. Some of the more common rabbitfishes include: the Double Barred Rabbit fish, the Java Rabbit Fish (*Top*), and the Gold Saddle Rabbit fish. Rabbit fish can be found solitarily, but more commonly will be observed in schools.

Importance

Rabbit fish are important reef herbivores feeding on benthic algae and biofilm. On many reefs in the Indo-Pacific they are one of the most abundant and important fish grazers. They are also an important prey species for many predatory fish including sharks, groupers, and barracudas. They are an important food source in many developing countries, and are threatened by over-fishing.

Notes

You do not need to differentiate rabbit fish by species or size for the EMP.

At night rabbit fish can be found sleeping with their dorsal and anal fin spines extended, making them a more difficult meal for nocturnal predators.

E. Snappers (Family: Lutjanidae)

Characteristics

Snappers are predatory fish with a **perciform** shape. They appear strong and streamlined when compared to the herbivorous fishes. There are several species of snapper in the Indo-Pacific, the most common in reef areas are of the genera *Lutjanus* such as the Russel's Snapper, the Spanish flag Snapper, and the Blackspot Snapper.

Importance

Snappers are **secondary or meso-predators** which feed on crustaceans and other fish. Some species also eat zooplankton, and are thought to be an important control on larval supply to the reef (Including COTs and Drupella). They are also a favorite of the fishing industry, and can be used to assess fishing related threats.

Notes

Snappers can be solitary or in schools, and are usually quick to flee from divers. They generally average 20-30cm, but occasionally much larger ones can be found on the reef.

F. Surgeonfish (Family: Acanthuridae)

Characteristics

Surgeonfish (also known as Unicorn fish) are mostly herbivores fishes, and derive their name by the very sharp spines protruding from the tail, which look like a surgeon's scalpel. They can be easily identified by their unique body shape and colorful spine (usually orange or yellow).

Importance

In some areas they tend to exist in large schools where they play an important role as a reef herbivore, often competing with damsel fishes for food. They are also prized in the aquarium industry and can be a victim of over-collection.

Notes

In many areas of the Indo-Pacific, Surgeonfish can exist in large schools, but in other areas they can be quite rare. In areas where they are rare, please record any Surgeon Fish outside the survey area into the notes section of your slate.

G. Sweetlips (Family: Haemulidae)

Characteristics

Sweetlips are part of the Emperor fish family, which can grow quite large. Their body shape appears like a mix between the groupers and snappers. Adults are generally white or dull colored with black spots, while juveniles are brightly colored and very ornate.

Importance

Sweetlips are important predatory fishes that feed on **crustaceans** and other benthic invertebrates. They are a favored fish by the fishing industry, and are indicators of fishing pressure on the reefs.

Notes

Juveniles and adults have very different appearances, and live in different areas of the reef. The juveniles can usually be found living solitarily in the shallow reef areas, and will stay in the same place for about 1 month while developing. Adults are more likely found in pairs or small groups, hiding under rocks or coral heads during the day and feeding at night.

H. Triggerfish (Family: Family Balistidae)

Characteristics

Triggerfish are large reef fishes with an oval body and a strong mouth. They swim using their dorsal and anal fins in an undulating motion. Many of the larger species are notoriously territorial, and can be quite aggressive (In particular the Titan and Yellow Margin Triggerfishes). However, some of the smaller species are herbivories, and are much more docile. They have two 'Triggers' which are used to warn intruders and lock themselves into cracks in the reef while sleeping.

Importance

Large Trigger fish are important top predators on the reef. They help to control the populations of coralivorous gastropods, mollusks, bivalves, and echinoderms (feeding on both Drupella and COTs). They are very intelligent fishes and often use tools such as rocks to break open clams or other shelled organisms.

Notes

Some species, such as the Titan Triggerfish, often attack when a diver enters their territory, especially during mating seasons. To avoid problems, do not swim directly towards or over them, but instead remain calm, lay low, and wait for the fish to move away from you.

I. Red Breasted Wrasse (Family: Labridae)

Characteristics

The *Labridae* family of fish (Wrasses) is very large and diverse. This species is one of the most widespread in the Indo-Pacific and larger Wrasse, growing up to 40 cm. They are easily identified by their black and white striped body and red head/breast. They can generally be found in areas of mixed coral and rubble.

Importance

Red Breasted Wrasse (also called Red Breasted Maori Wrasse) are large fish which require a high abundance of marine invertebrates to survive. Like the

Triggerfish, they have a diverse diet and are able to feed on hard shelled invertebrates and sea urchins.

NOTES

The Red Breasted Wrasse is often found in fishing cages, and is not easily frightened away by divers.

J. BLUESTREAK CLEANER WRASSE (FAMILY: LABRIDAE)

CHARACTERISTICS

The Bluestreak Cleaner Wrasse is a small (less than 10 cm) fish that generally stays in one place on the reef, its cleaning station. When potential 'clients' pass by, the fish will perform a sort of dance to indicate that it is open for business. If the client agrees, it will take on a passive body position (usually the head elevated and the mouth open. The cleaner wrasse will then remove ectoparasites from the client fish's body.

IMPORTANCE

Cleaner wrasse are important to the health and longevity of many other species of fish on the reef. Since they generally stay in the same area, we can accurately track their populations over time.

NOTES

There are several species of fang blennies which have evolved to imitate the cleaner wrasse, please learn the difference between the two so that you only count the wrasse.

J. MORAY EELS (FAMILY: MURAENIDAE)

Moray eels are actually a type of bony fish, they are not related to snakes as some may infer. There are over 200 species of moray eels worldwide. They have poor eyesight, but that is made up for by a highly acute sense of smell. They have two sets of jaws, oral and pharyngeal, the later of which can be articulated forward and back to rip apart prey.

CHARACTERISTICS

Moray eels have a **serpentine** shape, and live in burrows in the sand, or crevices and holes in corals and rocks. They feed on crustaceans, mollusks, and small fish. Most species are mesopredators, and sometimes become prey for barracuda, large groupers or sea snakes. Although species like the Giant Moray can be considered an apex predator.

IMPORTANCE

As predators, eels are important in regulating the balance of the reef. They are also sensitive to habitat destruction and declines in water quality. Some of the more rare and ornate Morays are also threatened by the aquarium trade, and can be sold for a high market price.

NOTES

They generally will not be found during very bright days, but can readily be found on deeper sites, when visibility is poor, or when it is overcast. This is one of the reasons why it is important to record the weather when conducting the EMP.

As you conduct the EMP survey you may notice that most of the fish that you actually observe are not indicator species. In order to progress your learning of the reef ecosystems, it is recommended that you also use the guide books to learn 1 new group of non-indicator species fish each EMP dive.

* * *

The next group of Chordates we will explore are the marine reptiles (Class: *Reptilia*). Both groups we will look at, the Sea Turtles and Sea Snakes, actually evolved from terrestrial ancestors. The Sea Turtles from land and freshwater turtles, and the Sea Snakes from Australian Elapids (venomous snake species) during the Jurassic period.

A. SEA TURTLES (SUPERFAMILY: CHELONIOIDEA)

Sea Turtles are a marine reptile, which first appear in the fossil records around 150 million years ago. They can live over a hundred years, and have very few natural predators after reaching adulthood. There are currently 7 species of sea turtles worldwide, and all are endangered or threatened.

CHARACTERISTICS

Of the 7 species of sea turtles, 5 can be found throughout the Indo-Pacific: the Hawksbill, Olive Ridley, Green, Leatherback, and the Loggerhead Sea Turtles. The Flat back is only found around the North coast of Australia, and the Kemp's Ridley is only found around the Gulf of Mexico.

IMPORTANCE

Turtles are an indicator of biodiversity and anthropogenic pressures. Sea Turtles have lived in the sea since the time of the dinosaurs, but today are faced with extinction due to human activities (primarily by marine debris, fishing, by-catch, and habitat destruction). As turtles migrate over great distances and through both national and international waters making their protection very difficult.

NOTES

If you are interested in learning more about Sea Turtles, Conservation Diver offers the 'Sea Turtle Ecology and Head-Starting Program' course. As most of our training centers have turtle nurseries, you can also assist in the care and maintenance of juvenile turtles. Ask your instructor for more information and how you can get involved.

B. SEA SNAKES (FAMILY: ELAPIDAE)

Sea Snakes are a marine reptile, which first appear in the fossil records around 130 million years ago. Most species live their lives entirely in the sea, and cannot move at all on land. A few species (Sea Kraits) are able to move around on land, but mostly only venture out of the sea to lay eggs.

CHARACTERISTICS

There are about 69 species of sea snakes, all within the Indo-Pacific, some of which are the most venomous snakes in the world. The have a paddle-like tail, and their lungs extend most of the length of their bodies so that they can hold air for

dives. They are generally found in shallow, warm tropical seas in sheltered areas or around islands. They generally feed on small fish, using chemosensory to detect prey as their vision is limited.

IMPORTANCE

Sea snakes are subject to overfishing and by-catch, with four species listed by the IUCN, 1 as vulnerable, 1 as endangered, and 2 as critically endangered.

NOTES

Although some of the most poisonous species of snakes on earth, Sea Snakes and Sea Kraits are generally not a threat to divers unless they are provoked. They are much more dangerous when on land (Sea Kraits) or when caught in a fishing net.

* * *

By the end of this chapter you should be able to identify all of the above species of fish used as indicators species in the EMP. Your instructor will help you to identify these fish underwater during the EMP, and also teach you about other fish species that are not included in the EMP. After completing the survey for fish you will then be ready to move to the most advanced survey of the program, the Substrate Survey, which is covered in the next chapter.

Chapter 4 Review

After completing the reading and discussion of the material covered in Chapter 4, you should understand and be able to answer the following questions. Please talk with your instructor about any questions you may have.

1. What is the primary threat to shark populations worldwide?

2. What is the primary threat to Sea Turtles populations world-wide?

3. How do we differentiate groupers during the EMP? Parrotfish?

4. Make a list of at least three species of fish (and other vertebrates from the chapter) for each of the following categories: Herbivore, Carnivore, Top-Predator.

5. Which of the indicator species are threatened by overfishing?

6. How can you tell the difference between a grouper and a snapper?

7. Which of the fish listed above are corallivores?

8. How many cubic meters is each survey section of the transect line for the fish survey?

Chapter 5: The Substrate Survey

"Learning the substrate survey will help you to view the reef in a more objective way, instead of seeing the reef as a single state, you will be able to infer the history and project the future state of the reef after only a single visit"

- Introduction to Substrates
- Substrate Types
  - Code
  - Description
  - Notes
- Review of Codes

Chapter 5: The Substrate Survey

One of the most important yet difficult surveys conducted during the EMP is the substrate survey. The word 'substrate' refers to the bottom composition on the reef, whether that be **Biotic** or **Abiotic**. For the substrate survey, participants will need to be able to identify living substrates such as corals, algae, and clams; as well abiotic reef components such as sands, silt, rocks, and rubble. After mastering the identification of substrate types, you will then assess coral growth forms and health of the colonies.

Remember that the substrate survey is done using a point intercept transect, whereby points are sampled under the transect line every 50 cm. This becomes complicated because on the reef there are over 9 million species of organisms, and many different types of inorganic components. We must lump many of these into groups to create a more manageable and simpler model of the reef, which means that you will often have to use good judgment and decision making skills to classify the substrate types. To make things faster we also use codes to classify the various substrate types, which is often the most difficult part for beginning students. Once you have mastered the classifications and coding, the substrate survey becomes much easier, and is often the preferred survey to perform by certified students of the program.

For this chapter, we will go through the substrate types and codes, and give some of the characteristics of each particular group. Next, we will look more closely at the category of Hard Corals, and assess both the growth forms and the health of the corals. Before beginning this section, it is recommended that you review the coral anatomy section found in Chapter 1 and the procedure for conducting the Substrate Survey in Chapter 2 of this manual.

Substrate types and Codes – Non-Living

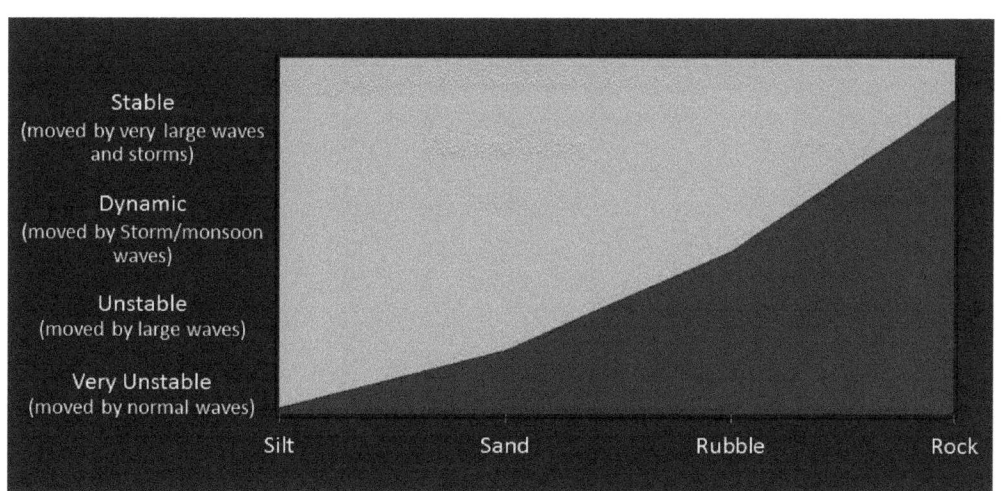

The non-living components on the reef are classified by their stability, or their potential to act as a surface for the recruitment of corals, sponges, and other marine organisms.

1. Silt/clay

Code: SI

Description

Silt is the smallest abiotic category; it consists of particles that are small enough to be suspended in the water column, that sometimes make up the sea bed. Generally, **SI** can be found on the reef flat, quiet lagoon areas, or in the deep sea areas. **SI** is not a common component of reefs, and if present in high abundance indicates disturbances such as erosion and sedimentation impacts from land. **SI** is easily disturbed by fins or waves, and remains suspended in the water column, often blocking out light and reducing visibility.

Notes

Silt is considered to be the most **dynamic substrate**, and corals will not be able to grow in silt laden areas. Areas which have silt inundation will quickly become less healthy and less diverse.

2. Sand

Code: SD

Description

Sand is any non-living particles on the seabed that is bigger than silt/clay but less than about 3cm in length/diameter. This is most commonly coral or silica sands as can be found on the beach, but also includes small pieces of dead shells and corals. Sand can be easily disturbed, but does not remain suspended in the water column.

Notes

SD is a dynamic substrate (although less so than **SI**), and corals which settle or fall into the sand generally do not survive.

3. Rubble

Code: RB

Description

Rubble is any non-living, non-secured substrate which is stable during normal conditions, but can be moved by seasonal storms and wave events. It consists of rocks, shells, dead coral or any other non-living components. Although less dynamic than SD

or SI, coral larvae settling in rubble are often killed later when the rubble is turned over by waves or feeding activities of fish.

NOTES

RB is a dynamic substrate, but allows for coral growth very slowly over time. Firstly, sponges and coralline algae grow between the rubble pieces, helping to lock them together. Secondly, corals growing on the rubble often are able to increase its size, before being killed during a large wave or storm event. Eventually new coral larvae settle upon and build up this enlarged rubble piece. Through this process, the size of the rubble can increase until it becomes stable enough to allow long-term coral growth, and thus a new reef. Often this process takes many decades to occur. Large fields of rubble often indicate some sort or previous disturbance of the reef such as bleaching, dynamite fishing, typhoons, etc.

4. ROCK

CODE: RC

DESCRIPTION

Rock is any non-living substrate larger than about 15-30 cm in diameter, and is usually more secure and stable (less dynamic) than the previous categories. Rocks can be both inorganic (granite boulders, cliff faces, etc) or organic (dead coral heads) in origin. **RC** is stable and provides good settlement points for corals and other sessile organisms, and often can attract a wide variety of marine life.

5. TRASH/RUBBISH

CODE: TR

DESCRIPTION

All man-made items (except artificial reefs) are coded as trash/rubbish. Some man-made items are not harmful to the marine environment but can actually provide habitat for animals (concrete, glass, etc.) However, any rubbish which is unsecured or moving (tires, nets, etc.) or toxic (batteries, plastic, etc.) is dangerous to the marine environment and should be removed.

Substrate Types and Codes – Living

For living substrates, we are concerned primarily with sessile organisms, those which after settling down do not move. These may provide for the living space or securement point for other organisms. Organisms which recruit onto other living animals are called epibionts, and include bivalves, polychaetas, algae, and many others. Any animal with epibionts should be considered a living substrate.

1. Sponges

Code: SP

Description

Sponges are the simplest marine animals, and are an important component of the reef. They can be identified by the **porous** outer tissue, and central vents or chimneys. Sponges are generally found growing on rocks, rubble, or in the sands. Sponges are very important in the cycling of nutrients on the reefs. Sponges often provide habitats for nudibranchs, sea cucumbers, bivalves, shrimp, and a wide range of micro-invertebrates.

Notes

Be sure not to confuse sponges with **tunicates** and **ascidians**, which use an in-current and ex-current siphon to filter water and have non-porous tissues.

2. Nutrient Indicator Algae

Code: NIA

Description

Nutrient Indicator Algae includes all of the macro-algae and the algal mats that smoother corals on the reef. It does not include the filamentous algae that covers all non-living substrates, including **TR**. Macro-Algae is in direct competition with corals, and reef areas with high abundance of NIA are generally considered unhealthy. Macro-algae can be both an indicator of excess nutrients in the water, or decreased herbivore activity by fish and invertebrates.

NOTES

Locations with the highest abundances of NIA tend to be those directly adjacent to villages, cities, or watersheds with high amounts of development. If you complete a dive and find an abnormally high amount of NIA, be sure to look around from the surface and try to identify any of the point-sources of erosion, development, or nutrient enrichment; and note these on you slate. Most often, once a coral reef shifts to an algal ecosystem there is little to no chance of coral recovery.

3. Soft Corals

Code: SC

Description

Soft corals, or **Octocorals**, for our purposes are any coral that, when dead, will not contribute to the structure of the reef. Generally, their polyps contain tentacles in multiples of 8. Instead of having a solid calcium carbonate skeleton, soft corals may have a hydrostatic skeleton, meaning that they inflate with water. Examples include the leather corals, finger corals, and brightly colored *Dendronephthya* corals. Other soft corals, such as the sea fans and sea whips, have a protienacious **skeleton**. Unlike hard corals, most soft corals often do not rely heavily on zooxanthallae, and can grow more quickly and in a wider range of environmental conditions than the hard corals (deeper/colder water, or in sandy areas). Because soft corals depend more on the zooplankton they can catch from passing water, they will often be found on submerged pinnacles or areas with strong currents.

Although soft corals do not contribute to the reef structure, they still provide habitat and food for a wide range of other marine species.

Notes

To tell if a coral is soft or hard, often you can wave your hand near the coral and look for any movement, if there is no movement due to the current created by waving your hand, the coral is most likely a hard coral.

4. Other

Code: OTH

Description

Any other sessile organisms living on the substrate. Note that this includes only non-moving organisms, for sea cucumbers or sea stars and other moving animals try to code what is below them, as that is what is likely to be there the next month when data is collected.

As many of the animals which fall into the other category may potentially be fouling organisms, those which compete with corals, it is important that you also note what was there at the point coded as **OTH**.

Tunicates can become a major problem for reefs where water quality has been altered, completely overgrowing and killing corals. They tend to thrive in more turbid, polluted, or nutrient rich waters. They have few major predators, and usually populations of tunicates are controlled by high wave action and strong currents which dislodge the animals. There will be annual fluctuations in their populations following heavy rains and other seasonal changes, but any areas that have persistent populations of tunicates should be closely monitored.

5. Hard Corals

Code: HC

Description

Hard Corals, or Scleractinians, are the reef building corals. They have a solid, calcium carbonate skeleton which provides both structure and habitat for other marine organisms. As discussed previously, they make up the foundation and are one of the most important class of organisms on the reef.

Notes

The reason that we have kept such an important group for last is that when you code a point as **HC**, there are several other pieces of information that you must code for. The first additional information needed is the shape, or **growth form** of the coral. Next, you also need to code for the health of the coral. These additional items are covered in the next sections.

Hard Coral Growth Forms

PLASTICITY OF CORAL GROWTH FORMS REPRESENTED THROUGH TYPES OF PLATY CORALS (NOTE, THESE PHOTOS ARE NOT FROM THE SAME CORAL COLONY/SPECIES, BUT ARE USED TO ILLUSTRATE THE CONCEPT)

Often, students of the EMP assume that, like the fish and invertebrates, they will have to learn coral species. Actually, learning coral species is very difficult, and takes years of learning and practice. Furthermore, the species of coral present does not tell us a great deal about the environmental conditions on the reef.

Recall that a single coral species can grow in a wide variety of shapes and sizes depending on the environmental conditions (light, currents, sedimentation, etc.). For example, we could take a funnel shaped coral growing at 10 meters depth and break it in half. We then take one half and move it to 5 meters depth, and the other half we move to 15 meters depth. If the coral does not die, then overtime the shape of the colonies will change. The shallow coral will be receiving much more light than before, and will grow in vertical, dense plates to create shade and reduce the amount of radiation hitting the colony. The half moved to 15 meters depth will receive much less light, and will grow flat like a plate to collect as much solar radiation as possible from sunrise to sunset.

Similarly, some growth forms will only occur in turbid waters, others where there is strong current, and some in only calm lagoons. Thus, assessing the shape of the coral gives us information on the conditions present on the reef. For these reasons, we will begin by learning the coral growth forms, which is the overall shape of the colony. If you continue further with your EMP training you will also learn the families and genera of corals, covered in Chapter 7 of this manual. As you go through the list of coral growth forms you will realize that they are in alphabetical order and clustered into groups (BCD, EFL, MS, T).

BRANCHING

CODE: B

DESCRIPTION

Branching corals grow in a tree like (**dendritic**) form. They have tapered, symmetrical branches with regular splitting. All of the polyps work together and contribute to the linear growth of the branch, which often has an **axial polyp** on the branch tip. They can be very fast growing (up to 10 cm/year), but are not very resistant to environmental stresses.

NOTES

Branching corals are structurally complex, meaning that they provide complex habitat for fishes and invertebrates, and often act as breeding or nursery areas for juveniles.

Branching corals are fast growing because they do not invest much energy into storage. As such, they have relatively thin tissues, and are susceptible to a wide range of threats. Branching corals are relatively weak structurally and are often broken or fragmented. If the broken branches fall into sand they often die, but if they fall onto rubble or a less dynamic substrate than they often can adhere and grow to create a 'new' colony. Through this form of asexual reproduction one single colony of branching coral can produce many colonies, all of which are clones (known as a **monospecific stand**).

Corymbose (or 'Crazy' Branching)

Code: C

Description

Corymbose corals grow in a bush-like form, with dense, irregular branches. Often it appears as each polyp begins to make their own branch, instead of working together to create a single branch as in the branching corals.

Notes

Dense branching leads to shading and reduced food availability for polyps growing on the inner branch, for this reason Corymbose corals are generally found in areas where light levels are high or currents/wave are strong. This means you will generally find them to be more abundant in the shallow reefs than the deep reef areas.

Digitate

Code: D

Description

Digitate refers to finger-like corals, which grow vertical, tapered, symmetrical branches which do not split. Many digitate corals eventually grow to become table corals when they are more mature. Digitate corals provide habitat for small fishes, most frequently of the Damsel or Butterfly fish families.

TABULATE

CODE: T

DESCRIPTION

Tabulate, or table corals, are generally made up of one of the three above growth forms (B, C, or D), which become tabulate in their more mature stages. Tabulate corals grow horizontally out over the sea floor from a central anchoring point, instead of growing vertically. They provide great habitat for juvenile fish or symbiotic crabs and shrimps, and often larger fish can be found under the tables.

As you may have noticed, coral growth forms **B, C, D**, and **T** are all closely related, collectively we could call this group the branching group. This group tends to be fast growing and provides diverse macro- and micro-habitats for invertebrates and fish. Being fast growing means that the skeletal structure of this group is not very dense or strong, and they are often broken by storms or careless divers and boats. Analogous to terrestrial fauna, these corals would be similar to weeds, which are advantageous growers that thrive when conditions are good but are unable to withstand disturbances or changes in the ecosystem.

The next group of coral growth forms are the thin platy corals, they tend to grow in thin sheets. Like the branching corals, they are closely linked alphabetically to make it easier for you to remember the coding.

ENCRUSTING

CODE: E

DESCRIPTION

Encrusting corals are thin, plate-like corals which grow along the existing substrate, like moss on a rock. They generally grow horizontally, and do not grow vertically. Their overall shape will match the existing substrate shape and topography that they encrust over. Encrusting corals will not only grow over rocks and rubble, but can sometimes be seen growing right over another living coral, killing it in the process (as in the picture at left).

In reality, all corals begin their life as encrusting corals before growing large enough to develop their own growth forms through skeletal development. However, these corals always retain that feature (not creating any vertical structure of their own) unless the reach an edge, in which case they may join the group called Laminar (explained below). Often, beginning students find it difficult to differentiate encrusting corals from other growth forms, but with a little practice this becomes much easier.

FOLIOSE

CODE: F

DESCRIPTION

Foliose refers to *foliage,* which is plate-like corals that often resemble flowers, lettuce, or other plants. This group includes any plate-like coral that grows in a 3-D shape, instead of flat along the substrate like the encrusting corals. Foliose corals which are flower-like are effective at collecting sunlight, but poor at dealing with sedimentation. Conversely, those growing like vertical plates are poor at collecting sunlight, but are nearly unaffected by sedimentation.

In fact, vertical platy corals like those on the last page (of the genus *Pavona*) actually grow like cups that collect sediment, lock it up through the assistance of coralline algae, and integrate it into their skeletal structure, allowing them to grow very fast. In shallow reef areas where sedimentation is a problem, these types of coral can outcompete most other species, leading to very abundant, but not very diverse, reefs.

Laminar

Code: L

Description

Laminar corals are thin, plate-like corals that grow horizontally, but free from the substrate, like shelves. Often an encrusting coral which extends over the edge of a rock or other substrate will become laminar. They are thin and delicate and will not withstand heavy waves or high diver/snorkeler impacts. They are also poor at shedding sediment. (Note: do not confuse laminar and tabulate corals, laminar are thin sheets, whereas tabulate corals are made of more robust branches)

* * *

After the coral growth forms E, F, and L, we come to the very dense and slow growing round or columnar corals. If branching corals where analogous to weeds on land, then the next groups would be the oak trees or bristle comb pines – slow growing, old, and able to withstand extreme environmental pressures. In fact, some corals within the next group can live for thousands of years. By drilling into the skeletal structures of these corals, scientists can look back into time (like rings on a tree) to get a view of ancient conditions on the earth, including climate, temperature, and disturbance events.

All of the growth forms we have looked at thus far have quite thin tissues as they do not store much energy in the form of lipids (fats). Instead, energy is primarily directed towards growth of the skeleton. The next groups we will explore, have much thicker tissues as a result of redirected energy towards storage for the future rather than short term growth. These corals are thus less susceptible to mortality by disease, or environmental factors such as bleaching.

Massive

Code: M

Description

Massive corals are slow growing (~1 cm/year), dense corals. They have a spherical or hemispherical shape, and are very robust (giving them the name massive). This includes the brain and lobe corals found on the reef. Massive corals can live for thousands of years and become several meters in diameter. Areas such as the shallow reef flats or deep reefs, where conditions are extremely variable or difficult to survive in, may often be dominated by massive corals. In areas where water and light conditions are very good, massive corals can often become overgrown by faster growing corals.

Sub-massive

Code: S

Description

Sub-massive corals are similarly dense and slow growing, but lack the rounded or hemispherical shape. Often, they are composed of bumpy ridges and valleys or have irregular and columnar shapes. Columnar corals should not be confused with branches or digitate corals (columns are dense and slow growing, and also tend to be non-symmetrical and non-tapered).

* * *

All of the coral growth forms we have investigated so far have been corals that attach to existing substrate and then begin to deposit skeleton to create their own growth form. All of these corals require stability and a secure basal attachment to grow. If they become unsecured, they can be turned over or broken and will face partial or complete mortality. The next group of corals are free-living, meaning that they do not require solid substrate on which to grow.

Solitary (Mushroom Corals)

Code: R

Description

Solitary, or mushroom, corals are non-attached, free living corals. Often the mushroom corals are a single polyp (non-colonial), the tentacles are located on the upper surface, and the slit in the center of the disk is the mouth. If a mushroom coral is turned over, it can draw water into its tissue and inflate like a balloon, allowing it to catch the next wave and be turned right side up again. They are often found on dynamic substrates such as rubble, and are one of the first species to colonize an area after a disturbance (called succession, or re-growth of an ecosystem). In many cases, reefs dominated by solitary corals are in the first stages of re-growth from a prior disturbance such as mass bleaching, typhoons, tsunamis, etc.

Summary Chart of Coral Growth Forms

B-Branching   C-Corymbose   D-Digitate   E-Encrusting   F-Foliate

L-Laminar   M-Massive   R-Solitary   S-Submassive   T-Tabulate

Coral health

After recording the growth form of the coral, we next need to assess the health. So in total, for hard corals we will have three codes to write down for each colony lying under the line ([1]HC [2]Growth Form [3]Health). The simplest way to assess coral health is by looking at the color of the coral. The intensity and color of the coral is a reflection of the symbiosis between the coral and zooxanthallae, the 'holobiant'. Generally, a darker coral is more productive and receiving more energy than a lighter colored coral, which has less zooxanthallae present. Although scientifically this is not accurate all of the time, and some corals are always pale or do not rely heavily on zooxanthallae, this is a good metric to use for beginning researchers. Other indicators of coral health include the metabolic rate, growth rates, **fecundity**, or budding rates, but these are all outside the scope of this training.

Healthy Coral

Code: H

Description

The coral colony has a dark, consistent color (or colors), is not broken or damaged, and appears to be growing (as opposed to receding.) In a normal year the vast majority of hard corals (>95%) will be coded as healthy. Often, we can tell the direction of coral growth by the white ridge or tip of the coral where the new skeleton is being laid down (look at the summary table for coral growth forms above for corals B, C, D, E, F, L, and T for an example of this).

Notes

Remember not to count tips or edges of corals as unhealthy. Often identifying coral health is difficult for beginning students, but your instructor can show you some examples of healthy and unhealthy corals while you are diving. Once you are familiar with the corals around the survey area it becomes much easier to identify any unhealthy colonies.

Partially Bleached Coral

Code: PBL

Description

Often the first sign of decreased coral health is a breakdown of the symbiosis between the coral and the zooxanthallae, resulting in a loss of color intensity and uniformity. This can be caused by a wide range of stresses ranging from disease to temperature and water quality to pollutants and sedimentation. Often when corals become partially bleached the threat is still present and active, as this is often a transient state for corals. Some corals may recover and become healthy again, while others may worsen or experience mortality. When you observe corals which are pale or have white spots you should also attempt to identify the cause, and note that on your slate. There are generally three types of **PBL** corals which you will not be required to know, but will make your data more useful if you can remember them:

Fully Bleached coral (living)

Ecological Monitoring Program Manual

CODE: FBL

DESCRIPTION

Fully bleached corals are greater than 90% white, due to a complete lack of zooxanthallae, but the coral animal is still living and intact. This is a stressful state for the coral to be in, and often leads to full or partial mortality. FBL corals are receiving only about 5-15% of their normal energy budget, and must rely on stored lipids and fats to survive. Corals which normally do not invest into lipid storage (branching and platy) will generally face much higher mortality rates during a bleaching event than those that do (massive, submassive, and mushroom corals). Fully Bleached Coral is a transient state and the cause of the stress is often still present. Be sure not to confuse FBL with dead coral, this means that when you look closely at the coral you will still be able to see the living polyps. Generally, when the coral is bleached it is relying only on predation to get their daily energy needs, so tentacles that are normally only extended at night will be extended throughout the day.

RECENTLY KILLED CORAL

CODE: RKC

DESCRIPTION

Recently Killed corals are identified by white skeleton with no living tissue. Most often, this is due to predation whereby the coral tissue is eaten and the skeleton is left behind. This can also occur after a bleaching event or other stress such as coral breaking and falling in the sand. Because the tissue has been removed very recently (within a few days to weeks) the coral skeleton has not yet been colonized by **filamentous algae** and will appear clean and white. Another major determinant is that the micro-structure of the corallites is still intact, and has not had time to be eroded mechanically or chemically. RKC is a very rare state for corals, and should not be confused with FBL or corals which have been dead a long time. When you see RKC corals during the EMP you should try to identify the threat, look around for predators such as the Crown of Thorns or *Drupella* snails, and look at the surrounding corals to see if there are any other colonies in a similar state.

NOTES

In the upper picture at left, we can see a large spot of bare white skeleton surrounded by healthy coral tissue. This is often indicative of chronic predation or coral disease. Even if the line ran on the healthy section, you would still want to note that

there was a problem with this coral. Notice that the center of the spot has a yellow coloration, as it becomes fouled by filamentous algae, from this we know that the mortality of this coral started from the center and has moved out in a radial fashion.

DEAD CORAL

CODE: DC

DESCRIPTION

The classification for dead coral is used for coral which has died recently (i.e. several weeks to several months), but has already been covered over in filamentous algae and has a dark grey to black color; it should not be confused with rubble and rock which is coral that has been dead for a long time (more than about 6 months to 1 year). In the case of dead coral, the calice (the skeletal cup that the coral polyp would normally sit inside) is still visible, and the shape of the coral colony is still intact. In the case of branching corals this would mean that the skeleton has not broken apart into rubble yet, and in the case of massive and sub-massive corals means bio-eroders have not yet bored into or changed the shape of the skeleton on a macro- or micro-level.

SUMMARY CHART OF CORAL HEALTH PROGRESSION

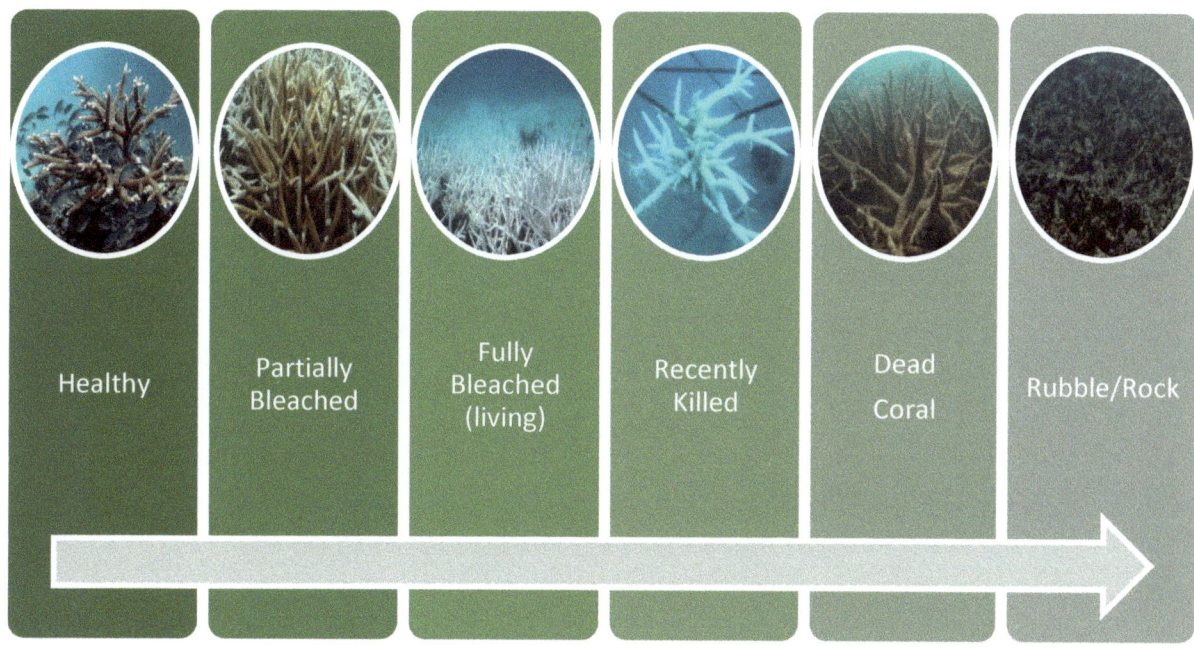

Review of the EMP Substrate Codes

Remember that for all substrates types except HC, there is only one code that needs to be written down. For HC, there are an additional 2 categories which must be assessed and coded.

Example Code: HC B H

Substrate Type	HC Growth Form	HC Health
• HC = Hard Coral	• B = Branching	• H = Healthy
• SC = Soft Coral	• C = Corymbose	• PBL = Partially Bleached
• SP = Sponge	• D = Digitate	• FBL = Fully Bleached
• NIA = Nutrient Indicator Algae	• E = Encrusting	• RKC = Recently Killed
• SI = Silt	• F = Foliose	• DC = Dead Coral
• SD = Sand	• L = Laminar	
• RB = Rubble	• M = Massive	
• RC = Rock	• R = Solitary	
• TR = Trash	• S = Submassive	
• OT = Other	• T = Tabulate	

* * *

Now you are familiar with all three of the main survey types used during the EMP program and can collect a full set of data. Remember that it will take time to perfect your skills, but with each survey you complete you will understand the ecosystem better and feel more confident in your monitoring abilities. Another essential way to improve your skills is to ask questions of your instructor, look up unknown species online, and check out the other books and learning resources that your training center will make available to you. The process of self-learning should never end. The next sections of the manual will help you plan the EMP and learn some of the more advanced techniques for those already confident in the chapters so far.

Chapter 5 Review

After completing the reading and discussion of the material covered in Chapter 5, you should understand and be able to answer the following questions. Please talk with your instructor about any questions you may have.

1. Explain how non-living substrates are classified based on their stability.

2. Label the chart below with the inorganic substrate types:

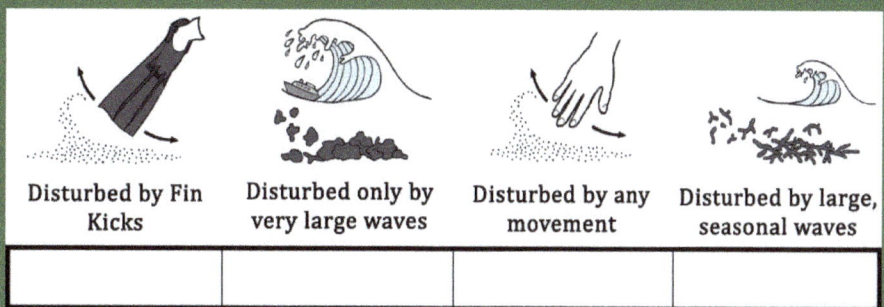

Disturbed by Fin Kicks	Disturbed only by very large waves	Disturbed by any movement	Disturbed by large, seasonal waves

3. Where would you expect to see the most NIA, shallow lines or deep lines?

4. How would you code a sea urchin under the line during the substrate survey?

5. Which of the coral growth forms do you suspect are the primary reef builders, and why?

6. Explain why some corals that are pale in color may still be coded as healthy.

7. When would you code dead coral skeleton as DC, and when would it be RB or RC, give some examples of the things you would look for.

Chapter 6: EMP Dive Planning

"The key to performing the EMP is good dive planning, logical thinking, and good judgment skills. Often changes to the plan will occur, and it is important that every member of the team knows how to react to those changes."

- Equipment
- Pre-dive planning
- Review of Codes
- Dive Procedure
- Debriefing and Data Entry

Chapter 6 – EMP Dive Planning

Conducting the EMP requires good dive planning, briefings, and efficiency. As you have learned, there is a lot to be accomplished on a single dive, and often if it is not all completed than the data from the dive will not be useable. This chapter is designed to help you plan your EMP data collection dive with 4 researchers (2 buddy teams) which is the suggested minimum. However, ideally you will have a group of 3 for each line. This plan can easily be adapted for more divers, but it is difficult to complete an entire data set with less than 4 people. During the EMP dive the following tasks must be accomplished by the research group:

- Locating, navigating, and laying out both transect lines (Shallow and Deep)
- Collecting full sets of data (4 segments on each line) for both lines for
  - Invertebrates
  - Fish
  - And Substrate
- Recording horizontal visibility using a secci disk
- Reeling in both lines and collecting all equipment

Required Equipment

A. **Student equipment**
   a. All personal standard equipment appropriate for the environment including:
      i. Mask, snorkel, and fins
      ii. Exposure suit (gloves are not allowed except in cases where a medical problem or low temperature warrants it)
      iii. Quick release weight belt or weight system
      iv. Regulator system with submersible pressure gauge
      v. Alternative air source suitable for sharing with other divers
      vi. BCD with low pressure inflator
   b. All equipment for performing the EMP dive
      i. EMP slate for fish, invertebrates, and substrate surveys
      ii. Pencil
      iii. Transect lines (4 x 50 m or 2 x 100 meter)
      iv. Compass (for the line navigator)

B. **Instructor Equipment**
   a. All personal, standard equipment as required by students
   b. Additional EMP equipment
      i. Extra pencil
      ii. Diving Knife
      iii. Secci Disk
   c. Safety equipment
      i. Surface marker buoy or float
      ii. Dive computer
      iii. First aid and oxygen supplies (on boat)

Ecological Monitoring Program Manual

Pre-Dive Planning

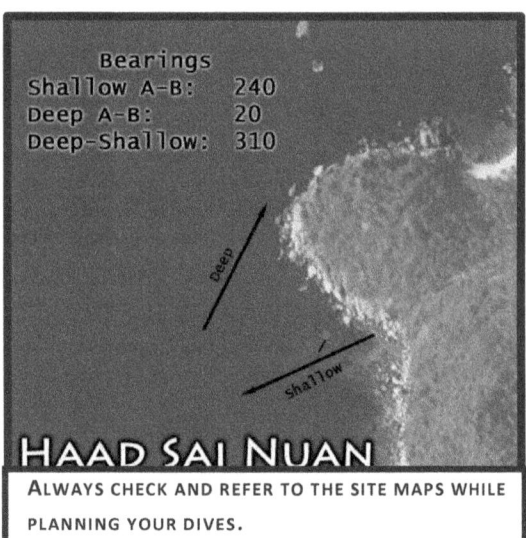

Always check and refer to the site maps while planning your dives.

Before beginning the dive, it should be decided which divers will conduct which surveys, and who will lie out and reel in the transect line. The group should then go through the dive plan (illustrated in the next section) to ensure that every diver knows the tasks they need to complete, since it is difficult to discuss this during the dive. Every diver should also know the tasks to be completed by the other teams, because if a team becomes low on air and has to ascend early, it is up to the other divers to take on additional responsibilities so that all the required tasks are still completed and all the equipment is brought back to the boat.

Other components of the dive briefing should include:

1. Site Background and History
    a. Site profile, depths of transect lines, types of reef ecosystem encountered
    b. Previous finding or disturbances in the particular site
    c. Any other relevant site information (i.e. Currents, waves, location of potential boat traffic, etc.)

2. Navigational bearings
    a. Bearing to Locate Deep A (plus any recognizable reef features to assist with search patterns if Deep A cannot be found)
    b. Bearing from Deep A to Deep B
    c. Bearing from Deep A to Shallow A
    d. Bearing from Shallow A to Shallow B
    e. Bearing to return to boat

3. EMP Dive Procedure for all divers (covered in detail in next section)

4. Low on air procedures
    a. Deciding to terminate the survey due to diver low on air, or give more responsibilities to other dive group
    b. Collecting the equipment if a diver becomes low on air
    c. Performing safety stop using Surface Marker and returning to the boat safely as a dive buddy team

5. Safety Precautions and Procedures
    a. Separation from the group or the transect area (lost diver procedures)
    b. Entanglement or other non-injury related emergencies
    c. Handling poisonous stings or bites
    d. Diver Recall Procedures

6. Proper Research diving skills and etiquette
    a. Maintaining Buoyancy while writing on the slate
    b. Ensuring safe distance from corals and substrate while laying out/reeling in the line
    c. Dealing with an entangled transect line
    d. 'Hands-Off' Policy for all divers

THE EMP DIVE PROCEDURE FOR 4 DIVERS

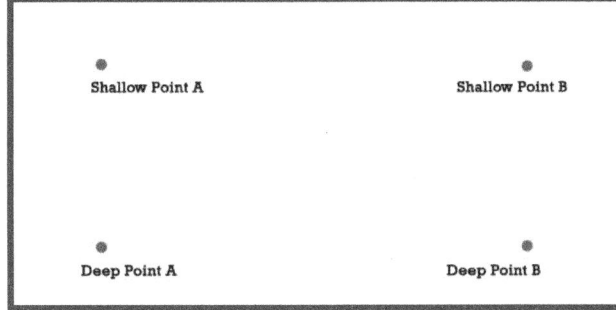

For the purposes of this example, the divers will be called 1, 2, 3, and 4. The buddy groups will be Group A (divers 1 & 2) and Group B (divers 3 & 4). Both groups enter the water at the same time and swim together to locate the Deep Line Starting Point (Deep Point A).

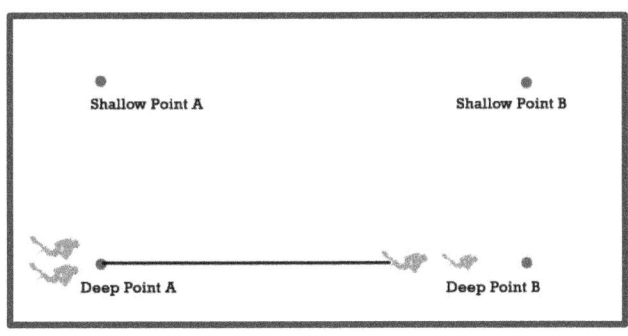

Next, diver 1 navigates the predetermined bearing while diver 2 lies out the transect line. Divers 3 & 4 perform the visibility check using the Secci disk. This is done by one diver remaining at 0m on the line holding the disk, while the other diver swims along the line looking back at the disk. When the disk is no longer visible, the diver marks down the position on the line, then swims back in slowly until the disk is visible again, the diver records this point on the line and averages the two to get the horizontal visibility.

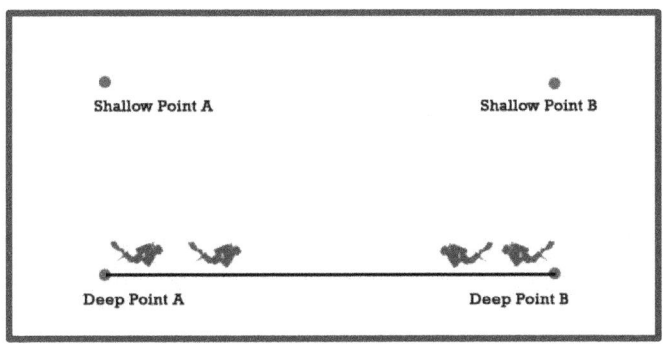

After waiting at least 5 minutes for the fish to return to the survey area, diver 3 then sets off down the line conducting the Fish Survey. Diver 4 follows behind conducting the survey for Invertebrates. Diver 1 returns down the line conducting the substrate survey, while diver 2 photographs any corals of interest or any substrate types which are unable to be identified by diver 1. When the two teams meet at the center of the transect line they should signal ok and also check each other's air supply so that any modifications to the dive plan can be made (i.e a diver is low on air and the other team must take on more responsibilities).

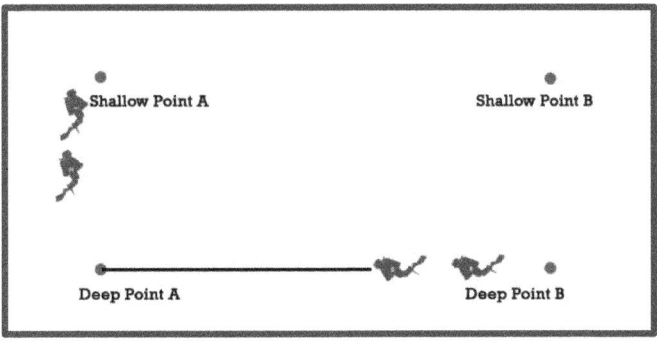

After completing the Substrate survey, divers 1 & 2 move to the shallow A point, while divers 3 & 4 reel in the line and collect any remaining equipment.

Alternatively, if there is only one dive leader, Group A waits for Group B to arrive before moving to the shallow line.

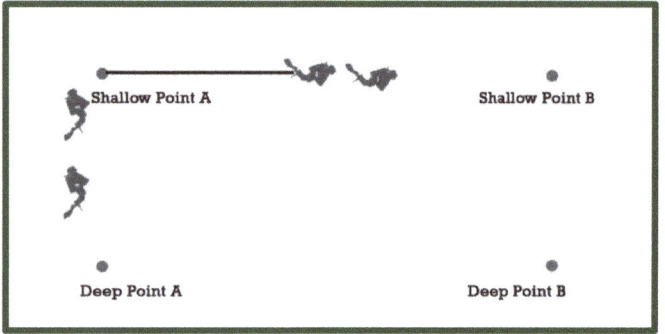

Diver 2 then navigates the bearing while diver 1 lays out the transect line. Divers 3 & 4 move to the shallow line and repeat the visibility measurement and wait a few minutes before starting the survey.

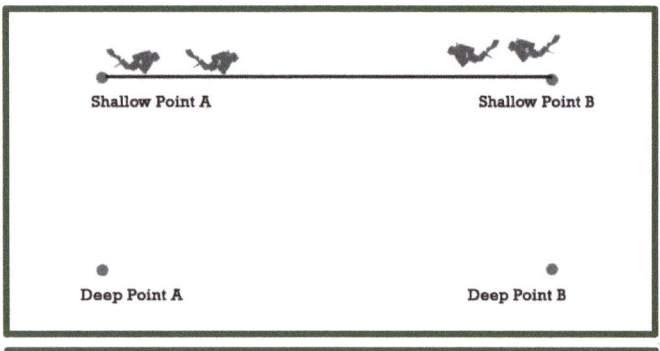

The divers switch tasks so that now diver 2 completes the substrate survey while diver 1 photographs; and diver 4 surveys for fish while diver 3 surveys for invertebrates. When the two teams meet at the center of the transect line they should signal ok and also check each other's air supply so that any modifications to the dive plan can be made (i.e a diver is low on air and the other team must take on more responsibilities).

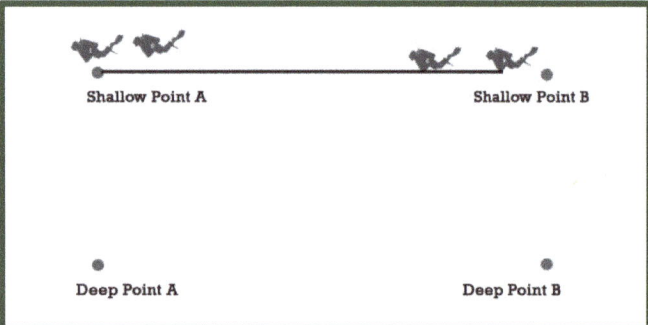

Upon reaching the Shallow A point, divers 1 & 2 collect any equipment and wait for divers 3 & 4 to return. Divers 3 & 4 reel in the transect line after completing the fish and invertebrate surveys. All of the divers return to the boat together with all of the equipment.

Post-dive Debriefing and Data Entry

After returning to the boat, the divers should discuss their impression of the reef, what they saw, and any interesting findings. Any of these points deemed significant should be recorded by the instructor and put into the EMP database or recorded in a journal/log-book. It is very important that this be done immediately upon returning to the boat or else vital information or pertinent questions can be lost or forgotten about. Students should also ask about any unknown species or growth forms observed during the dive while the information is still fresh in their minds, often this is more difficult upon returning to the dive school.

Upon returning to land, all of the data from the slates should be compiled, checked, and recorded onto paper. Data collected by the instructors and certified students should be recorded into the on-line database as soon as possible.

IMPORTANT NOTES

Only justified and accurate data should be put into the EMP database. The general rule is that if you are unsure of any aspect of the data, do not enter it into the database, as there is no way to go back and assess the data later on. Data to be entered must meet the following requirements:

- Both the A and B points of the line must have been found.
- Data must have been collected by an EMP instructor or certified student with at least 2-3 more dives of practice on the EMP beyond the initial course.
- Data must be complete (all 4 of the segments of the line must have been recorded).
- The data cannot contain any codes for unknown species, growth forms, or health.
- If visibility was less than 5 meters, fish data should not be entered, although invert and substrate are still allowed.

Chapter 6 Review

After completing the reading and discussion of the material covered in Chapter 6, try to plan an EMP dive. Remember to include the following:

1. Orientation to site map, and depth of transect lines
2. Locations of start points and bearings for both lines (and to travel between lines)
3. Who will navigate and lay out the lines (who will reel in the lines?)
4. Tasks each diver will complete on each line

Diver Name	Shallow Survey	Deep Survey
Ex: John Doe	Fish	Substrates
_____	_____	_____
_____	_____	_____
_____	_____	_____

5. Materials and equipment needed
6. Low on air and safety procedures

Ecological Monitoring Program Manual

Chapter 7: Coral Taxonomy

Chapter 7: Coral Taxonomy

In Chapter 5, you were introduced to identifying hard corals as part of the Substrate survey based on their growth forms. As mentioned, the growth form of a coral colony can be indicative of the ecological conditions and pressures of the area, as well as influence the types of marine invertebrates and fish that will be present. From an ecological standpoint, generally just identifying the corals according to growth forms can be enough for ecological surveys. However, now that you have already completed and hopefully mastered the 3 basic surveys for the EMP, it is time to begin learning some of the more advanced and in-depth survey techniques.

Introduction to coral families and Genera

Identifying the corals according to taxonomic descriptions (family, genus, and species) is an important part of any advanced coral reef monitoring program, and also a requirement for most biological or peer-reviewed research studies. Identifying corals to their taxonomic levels can provide important information such as:

- Biodiversity of corals in an area
- Changes in coral diversity over time
- Identifying target research species
- Monitoring threats such as predation, bleaching, or diseases; all of which will affect various species in different ways
- Monitoring corals for spawning and other advanced research or restoration projects
- Identifying rare or endangered species of coral
- Improving our understanding of coral population dynamics and changes

Not only is identifying corals important, many students also find it a challenging and rewarding process that gives them a better understanding of the reef. To date, over 1,604 coral species have been described, and it is estimated that there may be as many as 3,000 species on the planet today. No matter where you dive in the future, identifying coral species can easily become a lifelong passion for those who take the time to learn the basics of coral taxonomy.

For this section, you will be learning some of the major families of corals, and a few example genera within those families. These may not be the most abundant corals you have in your area, but as you learn them you will develop the tools for understanding the differences between types of coral using their visible morphological features. As you master this skill it will become much easier to learn new genera or species. This process does however take time, so don't get discouraged early, and just enjoy the process.

Selected Coral Families

Cnidaria
└→ Anthazoa
   └→ Sceleractinia
      Acroporidae
      Agariciidae
      Coscinaraeidae
      Dendrophylliidae
      Diploastreidae
      Euphylliidae
      Faviidae
      Fungiidae
      Lobophylliidae
      Merulinidae
      Montastraeidae
      Pocilloporidae
      Poritidae
      Psammocoridae
      Siderastreidae

Overview of Coral Taxonomy

As you will recall from previous chapters, corals are part of the phylum called Cnidaria, along with the hydroids (*Hydrozoa*), and jellyfish (*Scyphozoa and Cuboza*). Sea anemones, soft corals, corallimorphs, and zoanthids make up the class *Anthozoa* alongside the hard corals, which are known as the order *Scleractinia*. The order has gone under significant revisions over the last two decades as genetic analyses has become more accessible. As such, many of the species, genera, and even families have been completely revised.

The hard corals evolved some 535 million years ago. They almost went entirely extinct during the Permian extinction event 250 million years ago, but since then have diversified considerably. Today there are 29 families of corals in the order *Scleractinia*, containing more than 119 genera. A list of the more common hard coral families for the Indo-Pacific can be found at left. Because there are so many species, and identifying corals to the species level is very difficult, it often is enough to just identify and record corals to the genus level. Over the following chapter you will be introduced to the 20 most common genera on our reefs, and also gain the tools necessary to start identifying and recording some of the less common or rare groups.

How to identify coral genera

When identifying the taxonomic listing for any animal or plant species there are lists of anatomical parts to look for, known as the diagnostic features. Many of the characteristics which divers use to describe corals are actually non-diagnostic (color, size, shape, etc.) so first we must review the anatomy of a coral so that we know what features to look for.

Coral colony formation

The anatomical features of coral polyps were covered in Chapter 1, and it may be worth reviewing those before proceeding. Likely you have been paying more attention to the corals you are seeing on your dives after learning about coral anatomy. You have probably noticed that polyps come in many shapes and sizes. Sometimes the polyps are large and easy to see, while other times they may be compact and barely visible to the naked eye. While the growth form of a coral colony can change depending on the physical conditions, the polyps for any coral species will always grow in the same way. The shape and arrangement of the coral polyps is one of the first diagnostic tools we will use in identification. Reef scientists generally put coral polyp shape and arrangement into 8-10 categories, which are shown in the table on the next two pages.

It is not necessary that you remember all of the terms used to describe the colony formations, but it is important that you are aware of them, and know how to look for the shapes and growing methods of the corals while performing the EMP. As you look through the coral colony formation guides, you will probably start to notice that often these shapes are easier to see in the coral skeleton than the living coral itself. In fact, when identifying corals, the skeletal features are much more important than the living coral tissue. Generally, the features of the corallites are the most important features to look at for ID, which are essentially the area of the skeleton in which the coral polyps sit. In the next section we will look at these features in more detail.

Colony Formation in Corals

Solitary Attached - Single Polyp with distinct walls attached to surface

Solitary Unattached (Fungiod) - Single polyp, not attached to substrate

Plocoid - Polyps with distinct walls

Phaceloid - Polyps with distinct walls, growing in tubular or tree like pattern

Cerioid - Polyps with shared walls

 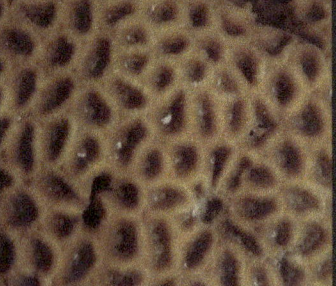 

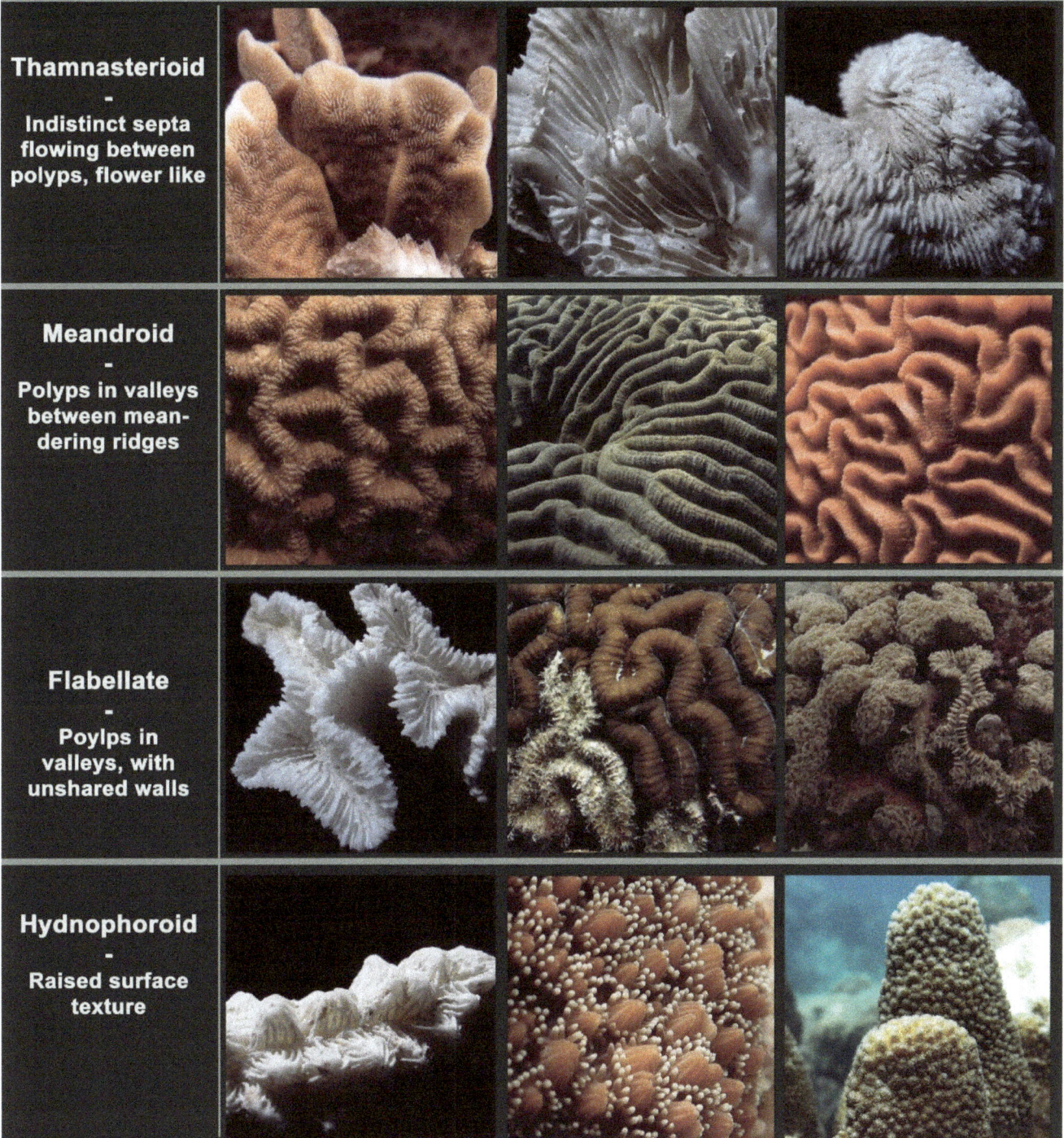

CORAL SKELETAL FEATURES

Several features of the corallites are going to be the most important features to us when identifying corals to the family or genus level. Although it is not always possible to see these features in the living corals, some of them will be apparent, and understanding these parts will allow you to use identification guides such as *The Coral Finder* (Russell Kelley 2009) or *The Corals of The World* (J.E.N. Veron, 2000) books to identify corals beyond the ones listed in this manual.

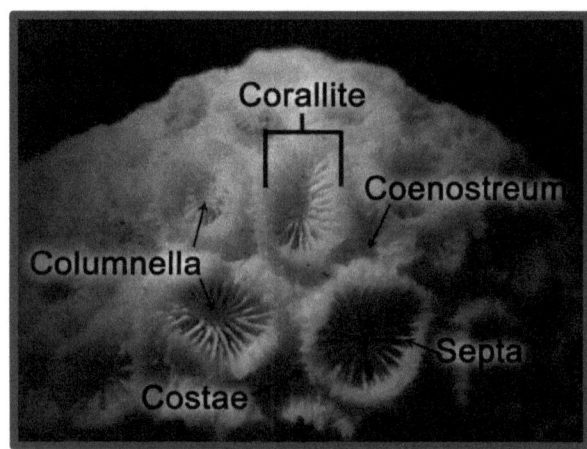

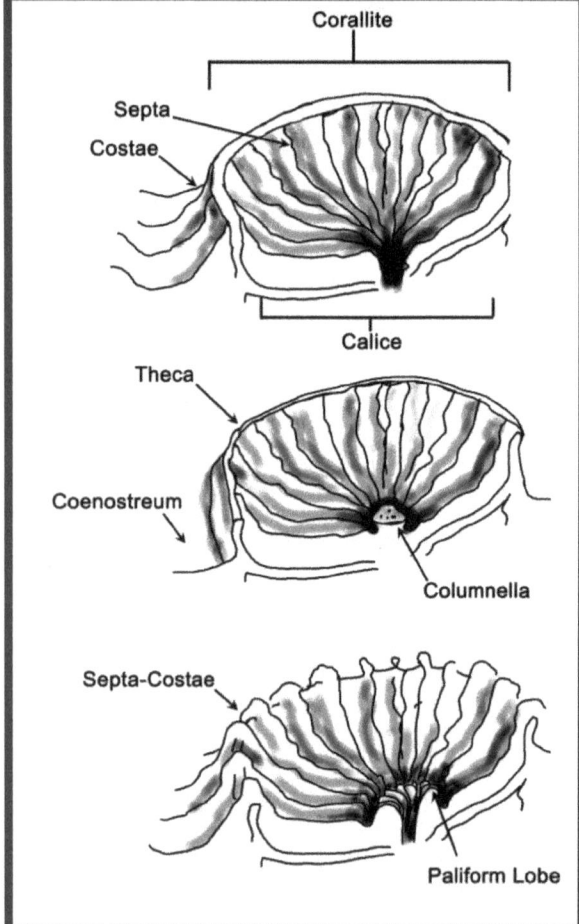

The corallite itself is the area where the coral polyps are, and is usually recessed down into the skeleton slightly. In a single coral colony, the corallites may differ in size, but structurally they are all identical (except in the genus *Acropora*, which will be discussed later). Generally, one corallite contains just one polyp, but in some species of corals there may be several individual polyps in each corallite. The recessed area of the corallite where each individual polyp sits is known as the **Calice**. Often the corallite will have a raised wall, which is known as the **Theca,** as you saw in the coral colony formation tables, these walls can be either shared (i.e. Ceriod, and Meandroid), separate (i.e. plocoid, phalecoid, and flabellate), or not present (Thamnasterioid, hydnophoriod, solitary unattached). In the case of meandroid corals, the calice is often referred to as 'valleys' and the theca as 'ridges.'

The area inside the corallite is usually full of vertical fins, known as **Septa**. Septa are one of the major features used for identifying corals to the species level, through counting and describing their cycles. For field studies such as ours, we will generally use more obvious features of the septa such as their size relative to the corallite and whether or not they have 'teeth'. In some corals, these vertical fins will also be present outside the corallite, in the region known as the **Coenostreum**; these fins are referred to as the Costae. If the Septa and Costae are continuous (such as in Thamnasterioid corals), they are referred to simply as **Septocostae**.

The center of each corallite or calice may contain a hole formed where the septa terminate, or there may be a feature rising up from the center known as a **Columnella**. In some cases the septa may dip down and rise up again to form what is known as a **Paliform Lobe**. From the top down the paliform lobe appears almost as another corallite within the larger corallite.

In some corals there may also be raised surface features in between the corallites, these are known as **Monticules** or **Verrucae**, and will be discussed in further detail in the descriptions for the genera of corals that they are relevant to.

Lastly, in identifying some corals such as *Montastrea* and *Favites* it is important to notice how new polyps are formed, called budding. The asexual reproduction of coral polyps is generally divided into two types; **Extra-Tentacular** and **Intra-Tentacular**. In extra-tentacular budding, new polyps are budded on the sides or spaces between existing polyps (outside of the

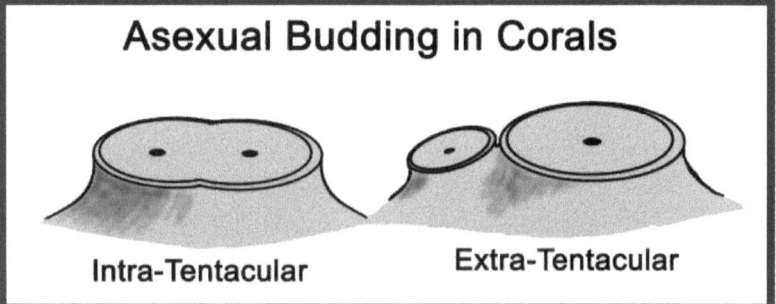

tentacular disk). Intra-tentacular budding occurs within the tentacular disk, and appears similar to cell division through mitosis.

IDENTIFYING CORALS FOR THE EMP SURVEY

As you may have already gathered, identifying corals is not always easy, and it may take several months or years to become proficient at it. So don't get discouraged, but instead pace yourself and try to learn one new genus from this book each morning and find it on the day's dive. The key to learning coral identification is practice, even if you memorize this section of the EMP Manual, you are going to find many corals during the surveys which aren't included or look very different from the examples provided.

As you learn to identify new corals, you can then record them during the EMP survey. For the EMP survey we will record only the genus level taxonomic names of the corals. To do so you will add the name as a code after the normal substrate codes, generally just writing the first 4-5 letters of the genus name (ie. **HC B H ACRO**). A full list of the coral genera and codes commonly used on Koh Tao can be found in Appendix E. These codes also correspond with those in the online EMP database.

In many cases you will not know the genus of the coral you are recording, and you can use the following techniques to help you identify it for the survey data or to learn it for next time:

- **Describe the diagnostic features of the coral**
    - What is the polyp arrangement?
    - How big are the polyps?
    - Do they have shared or separate walls?
    - What other features can you see (verrucae, monticules, ridges, etc.)? Try to describe the polyps/corallites in as much detail as possible.
    - Can you find any budding polyps? Record if they are Intra- or Extratentacular.
- **Take a guess at the family level**, many of the genera within a family share common features that can aid in later ID
- **Take a photograph**, or get another diver to photograph it for you. Be sure to get one picture of the overall colony and one close up of the polyps.
- **Draw a simple sketch** of some of the more unique features of the colony or corallites.

Remember to at least try and guess the genera or even the family, making some mark to indicate that you are unsure. Even if your guess is wrong, you will improve faster than if you simply skip all the corals you are unfamiliar with.

Common Coral Genera

Listed below are the 20 of the more common genera you are likely to encounter, the genera are grouped according to family, to aid your understanding of their relationships; however, it is not important that you memorize the family level information.

Family: Acroporidae

Acropora

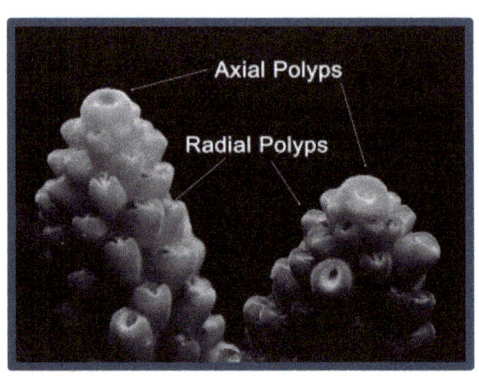

Acropora make up a large proportion of the branching, corymbose, digitate, and tabulate corals. *Acropora* are one of the most easily identified corals, due to fact that colonies have two different types of polyps, radial and axial. The axial polyp is called so as its growth is orientated in the same axis as the branch extension. The word *Acropora* is derived from the Latin words '*Akron*' meaning extremity or summit, and '*porous*' (pore). The corallites have indistinct septa, giving them a hollow 'cup-like' appearance. Generally, there are no costae or collumnella.

Acropora is listed as the main preferred prey species for *Drupella* snails, and also one of the preferred prey species for the Crown of Thorns Starfish. They are also a great habitat for small invertebrates and act as a nursery and habitat for larval or small fish.

Acropora is a diverse genus of coral, containing over 267 species distributed globally. They tend to be some of the fastest growing corals, extending their branches more than 5 centimeters per year. However, due to the fact they invest most energy into growth, they tend to be one of the less resilient corals on the reef. With less energy devoted to defense, mucus, and storage these species tend to be lost quickly during bleaching, disease, or predation outbreaks. Most of the *Acropora* corals in the Caribbean were lost due to disease outbreaks and related factors in the 1970's. In the 1998 and 2010 mass coral bleaching events, so much

Acropora coral was lost throughout the Indo-Pacific that some coral reef researchers think we are currently watching the slow extinction of this genus. Of the 22 coral species currently listed under the Endangered Species Act, 10 are from the genus *Acropora*.

ASTREOPORA

Corals of the genus *Astreopora* tend to be either massive, encrusting, or laminar. The polyps of *Astreopora* look similar to those of Acropora, but there is only 1 polyp type in the colony (no axial or radial polyps). Similarly to *Acropora*, these colonies tend to be relatively fast growing and susceptible to bleaching, predation, and other threats.

There are about 28 species of *Astreopora*, and often they are referred to as star corals due to the appearance of the corallites. Corallites tend to be conical, large (up to 3-5 mm) and distinct. The surface of the colony appears rough due to tiny bumps that cover the coenostreum.

Montipora

Coral of the genus *Montipora* tend to be encrusting, laminar, foliose, or submassive. On a macro-level, the polyps appear quite different to either *Acropora* or *Astreopora*. The corallites tend to be very small (less than 2mm) and relatively far spaced apart. The coenostreum may be covered in large bumps of skeleton called monticules or verrucae. Often, students comment that *Montipora* corals are difficult to distinguish from *Porites*, as the polyps are of similar size. The main differences are (1) that *Montipora* polyps tend to be widely spaced compared with the tightly packed *Porites* (ceriod); and (2) the tentacles of *Montipora* polyps tend to face outwards from the corallite, while *Porites* point inwards.

The genus *Montipora* is very diverse, with about 177 recorded species. *Montipora* corals tend to be slightly more resilient than *Acropora*. Often in shallow seas they can be observed encrusting over other corals or sponges.

Family: Agariciidae

The family Agariciidae contains about 25 extant genera and over 48 species. They generally have thin tentacles which can be difficult to see, and the corallites generally lacks raised walls.

Pavona

Pavona corals are generally foliose, and occasionally encrusting.

Colonies are made of interconnected fronds or leaves, generally orientated vertically. There are polyps on both sides of the plates, unlike foliose *Turbinaria* or *Montipora;* which only have polyps on the top surfaces of the plates. The polyps are small (less than 3mm), Thamnasterioid and lack a theca. Septocostae are very fine, and can only be observed close up.

There are about 22 described species of *Pavona*, one of which is listed on the Endangered Species Act. *Pavona* corals have a high propensity for fragmentation and tend to be a resilient coral - out-competing other corals and becoming dominant in areas under chronic stresses such as sedimentation or over-use. As such, *Pavona* can be used as an indicator for chronic stresses to the reef areas.

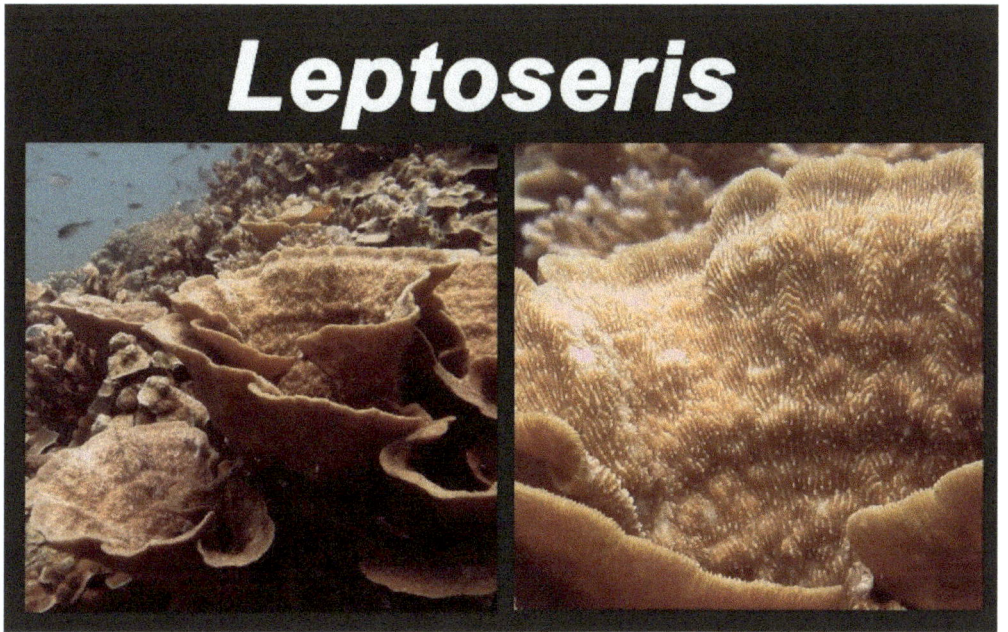

LEPTOSERIS

Leptoseris is another species of platy coral that tend to be foliose, laminar, or encrusting. Corallites have poorly defined walls, a collumnella, and fine septa-costae. Unlike *Pavona*, the polyps grown on only 1 side of the plates.

There are about 18 species of *Leptoseris*. They can be found on the shallow reef, as well as in some of the deepest reefs (up to 60 meters).

FAMILY: DENDROPHYLLIDAE

Most of the corals in the family Dendrophyllidae tend to be azooxanthellate, meaning they do not contain algae symbionts. These means that the are mostly heterotrophic, and tend to have large and prominent tentacles that are extended both day and night. Currently there are 169 listed species in the family.

TURBINARIA

Corals of the genus *Turbinaria* tend to be encrusting, laminar, or submassive. Corallites tend to be relatively large, round, and distinguished or widely spaced. The tentacles tend to be relatively long for the polyp size and often extended in the day time, and the coenostreum is usually quite smooth. *Turbinaria* are one of the few genera within the family *Dendrophyllidae* which do contain zooxanthallae.

There are 13 species of *Turbinaria*, distributed throughout the Indo-Pacific. *Turbinaria* tends to be easy to identify in the field, however students occasionally confuse it with *Astreopora*. While the colonies can look similar, on closer inspection they are easy to distinguish as *Turbinaria* has smooth, widely spaced polyps; and *Astreopora* has bumpy, tightly packed corallites.

TUBASTRAEA

Corals of the genus *Tubastraea* form either branching colonies or small clusters (similar to digitate). The polyps are phaceloid and relatively large and round. *Tubastraea* is a non-reef building coral, as the skeleton it secretes is thin and brittle. It is also an azooxanthellate coral, meaning that it is completely predatory and does not harbor photosynthetic algae. As such, color is diagnostic, with colonies being green, yellow, or orange.

There are just 7 species of *Tubastraea*, generally growing in deep (up to 110 meters) or turbid waters and will not be found in shallow reefs, except under rocks or overhangs.

FAMILY: DIPLOASTREIDAE

The Family Diploastreidae is monotypic, meaning it contains only one genus, and in fact contains only one extant species.

DIPLOASTREA HELIOPORA

Diploastrea heliopora is almost always massive, and can grow to several meters in diameter. The polyps are large (1-1.5 cm), conical, compact, plocoid, and regularly spaced; giving this coral the nickname 'Honeycomb Coral'. The skeleton is extremely dense.

The colonies are almost always brown, and are easy to identify and distinguish from other genera. They are widely distributed throughout the Indo-Pacific.

FAMILY: EUPHYLLIIDAE

The family Euphylliidae contains 7 genera and just 23 species. The name means "true leaf" and refers to the large septa of many of the species within this group.

GALAXEA

Corals of the genus *Galaxea* tend to be encrusting or submassive, sometimes creating large columns. The corallites are cylindrical and have obvious large protruding septa that rise high above the theca and polyp tentacles. The polyps are generally out in the day time, and are occasionally so dense and long as to occlude the large septa. Colonies tend to be brightly and ornately colored. There are about 8 species of *Galaxea*, which generally grow to depth of about 20 meters. They are known to grow well in more turbid waters, as they feed both day and night. They have long sweeper tentacles which are used to defend their space against any competing corals.

FAMILY: FAVIDAE

Favidae used to be one of the largest families in terms of genera, with nearly as many species as *Acroporidae*. However, recent revisions have reduced the family size greatly. Today the family contains only 22 species. It is also one of the more difficult families of coral to identify, as many genera look similar or can take many different forms depending on the species.

FAVIA

Corals of the genus *Favia* are almost always massive, but may sometimes be encrusting. The corallites are similar to the size (8-12mm) and appearance of *Montastrea* and *Favites*, except they are plocoid and the septa/costae do not alternate in size or thickness. They are also distinguished from *Montastrea* by having intra-tentacular budding (black arrow). The genus *Favia* contains 3 species.

FAMILY: FUNGIIDAE

The family *Fungiidae* is very distinct from all other families of coral, and contains 53 species. Most of the genera in the family *Fungiidae* are free-living (unattached) and Solitary (have only one mouth). *Fungiidae* corals are some of the oldest found in the fossil record, and it is thought that many of the colonial species of coral evolved from them.

FUNGIA

The genus *Fungia* used to contain about 30 species, but recent revisions have determined It to be monotypic, contain only 1 species, *Fungia fungites*. They are almost always unattached or free living as adults, although larvae usually recruit onto a solid surface. *Fungia* are always solitary, being just a single polyp/corallite.

Fungia are resilient corals, and are considered semi-mobile, able to right themselves when overturned and unbury themselves from sand. They are also a pioneer species; one of the first to recolonize a disturbed area. They play an important role in adding structure to areas that have been disturbed by physical damages, and provide a favorable growing habitat for coral recruits.

FAMILY: LOBOPHYLLIIDAE

Lobophylliidae is a recently revised family which contains 11 genera and 58 species.

LOBOPHYLLIA

Corals of the genus *Lobophyllia* are almost always massive, and can grow to more than 1 meter in diameter. The large (2-4cm) corallites are phaceloid or flabellate, but often packed closely together as to look meandroid. The corallites have a columnella and large septa which are covered in spikey teeth that sometimes protrude through the tissue along the theca. There are about 10 species of *Lobophyllia*.

FAMILY: MERULINIDAE

The family Merulinidae is a large family of corals, currently containing 148 species. Many of these species were previously listed under Faviidae until recent revisions. The two families are closely related, and often species are difficult to differentiate during field surveys. Unlike Faviidae, polyps of species within Merulinidae are generally highly fused and lack paliform lobes.

FAVITES

Corals of the genus *Favites* tend to be massive, submassive, or encrusting. Corallites are ceriod with common walls and range in size from 3-15mm. Septa have teeth, and give many of the species a rough or ragged looking texture, especially along the theca. The genus contains about 14 species. Often colonies can be bicolored. Skeletons are dense, and prolific in the fossil record (dating back to about 160 million years ago)

GONIASTREA

Corals of the genus *Goniastrea* tend to be massive, submassive, or encrusting. Corallites may be ceriod or meandering, and range in size from 3-15mm. Corallites usually lack a columnella, and often have a well-defined paliform lobe. The septa tend to be fine, neat, and smooth, and the corallites tend to be deep or have a 'dug out' look.

There are about 10 species of *Goniastrea*, with a wide range of corallites sizes and arrangements; and they are often confused with other species such as *Platygyra* or *Favites*. The main distinguishing feature however is the deep corallites, paliform lobe, and neat appearance of the septa. *Goniastrea* is a resilient and hardy species, and often dominates the shallow reef flats.

HYDNOPHORA

Corals of the genus *Hydnophora* tend to be submassive or encrusting. They have a unique appearance, due to the conical mounds, or hydnophores, rising from the center of the corallites. The tentacles of the polyps are arranged around the base of the hydnophores, with one tentacle between each set of septa. There are currently 6 species of *Hydnophora*.

PECTINIA

Corals of the genus *Pectinia* are usually foliose or laminar. Their name means 'comb-like', and refers to large and prominent septocostae. Corallites are large, thamnasterioid, lack a wall, well-spaced, and irregularly arranged. *Pectinia* is not a major reef building coral. This unique coral should not be confused with *Pavona*. There are about 8 species of *Pectinia*.

PLATYGYRA

Corals of the genus *Platygyra* form massive or submassive colonies. Corallites range in size from 4-10mm wide and tend to be meandroid, but occasionally may appear more ceriod. Compared to *Goniastrea*, the corallites and septa are not as neat and regular. Corallites generally have an indistinct columnella and do not have a paliform lobe. They are also a known host of the parasitic copepod *Bradypontius pichoni*. There are about 12 species of *Platygyra*.

MONTASTRAEIDAE

MONTASTRAEA

Corals of the genus *Montastrea* tend to be massive or submassive. The corallites tend to be large but varied in size (4-15 mm), irregularly shaped/sized and crowded. The septa alternate in size or thickness, and new corallites are produced through extratentacular budding. *Montastrea* was previously part of the family *Faviidae*, until it became its own monotypic genera. It however closely related, and easily confused with the genera *Favia* (which has intratentacular budding)

FAMILY: POCILLIPORIDAE

The family Pocilliporidae contains 4 genera comprised of 54 species, all of which are polymorphic, meaning the colony shape is highly carriable depending on conditions. They tend to have small corallites, prominent columellae. The group is closely related to Acroporiidae.

POCILLOPORA

Corals of the genus *Pocillopora* almost always form corymbose colonies with dense branches. The colonies tend to be lightly colored, usually brown or yellow. The corallites are very small (1-2mm), and simple. There are no visible septa, theca, or costae. The coenostreum tends to be smooth, but the colony surface is covered in large knobs or bumps called verrucae. Unlike *Montipora*, the bumps are covered in polyps; and there are no axial polyps as in *Acropora*.

Ecological Monitoring Program Manual

There are about 22 species of *Pocillopora*. One species in particular (*P. damicornus*) is considered to be a 'weedy' coral. It is prolific in reproduction and recruitment, and also has a high propensity for fragmentation, making it an effective pioneer species. Often it will out-compete other corals during periods of desirable conditions; like *Acropora*, devoting most energy to growth and not storage or defense. High rates of mortality amongst *Pocillopora* during disturbance events are common, leaving behind low density coral skeletons which quickly breaks down to leave fields of rubble.

FAMILY: PORTIDAE

The family Portidae contains 4 genera consisting of about 99 species. The family is not closely related to any other families, and the genera are quite heterogeneous. The coenosteum is quite small and the septa and walls are porous.

PORITES

Corals of the genus *Porites* primarily form submassive colonies, but can also be encrusting or arborescent (branching or tree-like). The corallites are very small (<1.5mm) ceriod, and have well defined common walls. Because the corallites are very small and the tentacles short or retracted in the day time; the overall colony appears quite smooth.

Currently, there are about 57 described species of *Porites*. There are widely distributed and occur across a broad range of habitats and depths. In many areas they are one of the primary reef building corals and also some of the oldest corals, reaching sizes greater than 4 meters in diameter. They are a resilient coral, and recover well from bleaching, predation, and structural disturbances.

GONIOPORA

Corals of the genus *Goniopora* tend to be submassive and columnar. Living colonies are easily identified by the large day-time extended polyps, usually purple or red in color. The corallites of the non-living skeleton resembles that of *Porites*, being ceriod with distinctive shared walls. Visually they are very similar to *Alveopora*, but can be differentiated by the number of tentacles; *Goniopora* having 24, and *Alveopora* having 12. There are about 27 species of *Goniopora*.

* * *

By now you may be feeling a bit overwhelmed by all these new taxonomic names and terms, but don't be discouraged. After reading this chapter you should have the knowledge and tools you need to begin developing your skills of coral identification. Don't forget to also check out the Coral Finder and Corals of the World books to for more information on these genera as well as the many other that occur in our area. There is a list of the codes for the most common genera around the island in Appendix E, try to learn 1 each day and you will be a master of coral ID in no time.

Chapter 7 Review

After completing the reading and discussion of the material covered in Chapter 7, you should understand and be able to complete the following exercises. Please talk with your instructor about any questions you may have.

1. Use this manual and the Coral Finder to identify at least 8 corals either from a book or collection your instructor provides you

2. Write a description of three corals from the collection using the terms listed in this chapter to describe colony formation, budding, and the parts of the corallite.

3. Identify at least 3 genera of coral on your next EMP substrate survey.

Chapter 8: Intro to Surveying for Coral Diseases

Chapter 8: Surveying for Coral Diseases

Disease affects all populations of animals and plants, and has always played a major role in the dynamics and evolution of the species living on our planet. However, climate change, pollution, habitat destruction, and over-use have increased the frequency and severity of disease outbreaks in many coral reefs over the last 4 decades. Whereas coral disease was barley recognized as late as the 1960's, today it has risen to one of the leading causes for coral reef mortality. In this chapter you will be introduced to coral diseases and how to assess corals during the EMP survey for some of the most common or destructive known diseases.

Introduction to coral diseases

In the early 1970's many of the reefs around the Caribbean were covered in dense stands of *Acropora* corals, known locally as Staghorn and Elkhorn coral. At that time the average coral coverage on the reefs was about 70%, and to most people in the area it seemed as though no matter what happened on land, those ancient reefs would persist forever. Then, in the 1970's the world's first recorded outbreak of coral disease struck. Known as White Band, this infectious disease almost completely wiped out all of the *Acropora* corals in the region. After two decades and the addition of several other problems, corals reefs in the Caribbean were reduced to an average of about 5% coral cover, and both species of Caribbean *Acropora* were added to the Endangered Species Act.

This devastating event took the coral reef world by surprise, as very few reef scientists had ever studied diseases, and there were no protocols for doing so.

Subsequently, outbreaks of other diseases have been recorded in the Great Barrier Reef, the Red Sea, and Hawaii. Today, the field is still struggling to catch up. More than 30 coral diseases have been described and recorded, yet the pathogens responsible are only known for about 9 of those. Furthermore, few reef scientists and even fewer reef managers are trained to recognize or investigate disease, and nobody knows what to do to stop them. There is much we don't know about coral diseases, which makes the regular monitoring and data collection even more vital. The groundwork we lay down now will hopefully lead to management possibilities in the future.

Luckily, we don't have to understand everything about coral diseases in order to study and monitor for them. In fact, you probably already know a lot more about diseases than you think, just from the experiences you have had with them during your lifetime. For example, you are probably aware that disease in humans can be caused by viruses, bacteria, and fungi; and the same goes for corals. Almost intuitively we know that in hot and unclean environments like an urban slum we are likely to get sick from bacteria; and again we can expect to

see similar chains of cause and effect in corals. In the reefs closest to human population centers is where diseases are taking the largest toll on coral health and abundance, and as global populations continue to increase, this effect is only becoming more exaggerated.

On the island of Koh Tao, Thailand, a recent study by Lamb *et al.* (2014) found that in 'low use' sites the prevalence of disease was only about 5.2%, but in 'high use' sites that rate increased to about 14.5%. A similar study on the Great Barrier Reef (Sisney et al. 2018) found that in general use areas all disease were higher than in protected zones (13.14% vs 2.37%)

Before you head out to start finding diseases, we must first review some of the basics about diseases and the terminology used to describe them.

Coral disease basics and terminology

By the term coral disease, we are referring to any abnormal condition, with describable signs and symptoms, which negatively affects the health and reproduction of the animal. In other words, we are going to be looking at anything that harms the living coral. Diseases caused by some pathogen or parasites (viruses, bacteria, fungi, etc.) are known as **Biotic Diseases**. Those caused by physical conditions (temperature, UV radiation, heavy metals, etc.) or toxic chemicals are known as **Abiotic Diseases**.

Some diseases can be transferred from an infected host to other members of the population, and these are known as **Infectious Diseases**. Other diseases cannot be transmitted, and are known as **Non-infectious Diseases**.

Many diseases can be described and classified as a combination of the categories above. For example, lung cancer is generally a non-infectious abiotic disease, as it is caused by smoking and air pollution and cannot be transmitted from one person to another. The flu on the other hand would be an infectious biotic disease, as it is caused by a virus that does move from person to person. As an exercise, try to think of a few other human examples of different disease types.

When describing diseases, three terms are generally used; Prevalence rate, Incidence rate, and Mortality rate. The **Prevalence rate** describes the number of individuals affected *at any one time*, it is usually described as a percentage and obtained from the equation:

$$\frac{Number\ of\ affected\ individuals\ in\ the\ population}{Total\ number\ of\ individuals\ in\ the\ popualtion} \times 100$$

The **Incidence rate** is the number of new infections or affected individuals over a given period of time, described by the equation:

$$\frac{Number\ of\ new\ infections\ over\ the\ time\ period}{Size\ of\ the\ at\ risk\ population\ over\ the\ time\ period}$$

Mortality rate is the number of individuals that died in a population over an amount of time, and is given by the following equation:

$$\frac{Number\ of\ deaths}{Total\ popualtion\ size} \times \frac{1}{time\ period}$$

Another important factor to recall is **Disease Triggers**, or the factors that can contribute to disease. In humans, we may think of the virus that gives us the flu, the temperature change that gave us a cold, the UV radiation that caused skin cancer, or the high sugar diet that caused diabetes. In corals, there are many different triggers for disease that include:

- Water quality effects due to run-off, sedimentation, and increased nutrient inputs
- Increased water temperature due to thermal anomalies and climate change
- Physical damage and stress
- Marine and terrestrial based pollutants and toxic chemicals, especially those from the use of chemical fertilizers and pesticides
- Transmission from fish farming and aquaculture
- Ocean acidification due to increased atmospheric CO_2 levels
- Ingestion of marine debris such as micro-plastics

There are also many ways in which infectious diseases can move from one coral to another, known as the vector of transmission. Some diseases, especially bacterial ones, can be transmitted from one colony to another via the water column. In viral diseases however this is less likely, due to the immense distance between coral colonies relative to the size of the virus. In these cases, coral predators often play a role in the transmission of disease. Both the Crown of Thorns Starfish and *Drupella* snails have been identified as vectors of transmission for several coral diseases, and so have butterfly fish.

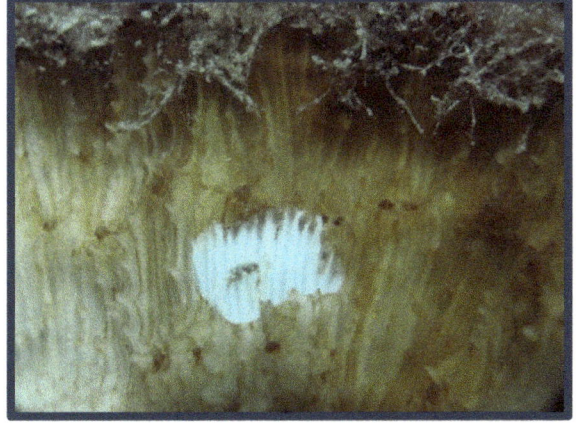

Diseases in corals are generally identified by discoloration, tissue loss, or the appearance of colored bands. Scars, or areas of tissue loss, are known as **lesions.** Lesions are described in several ways based on their (1) patterns, (2) rate of progression, (3) color, (4) thickness, (5) shape, and (6) border. When you learn to identify coral diseases, or when recording them in the field you will need to use specific terms relating to these characteristics.

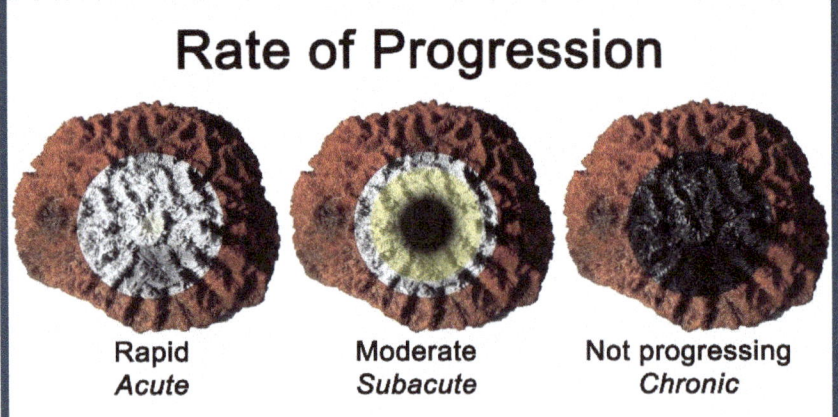

(1) The pattern of the lesion is generally described in 3 ways:
- **Focal:** Beginning from one point and spreading outwards
- **Multifocal:** Beginning from multiple points, or spotted
- **Diffuse:** Not having an obvious center, occurring randomly

(2) The rate of progression describes how quickly the lesion is spreading, and can be deduced by the amount of white, recently killed coral showing. After a few days to a week, the white skeleton will be colonized by filamentous algae and biofilms, taking on a yellow color. Once the algae and biofilm are completely developed the skeleton will appear dark grey or black.

- Rapid or **Acute**: The lesion is mostly or completely white, indicating both rapid and ongoing progression.
- Moderate or **Subacute**: The lesion shows various banded stages of clean skeleton and the colonization of filamentous algae.
- Not Progressing, or **Chronic**: The lesion is completely colonized by well-developed filamentous algae and biofilm, indicating that progression has ceased or is at a rate lower than a few mm per month.

(3) The lesion color is simply a description of the color of the afflicted area.

(4) The thickness refers to the margins of the lesion or banded diseases, and is usually given in mm or cm.

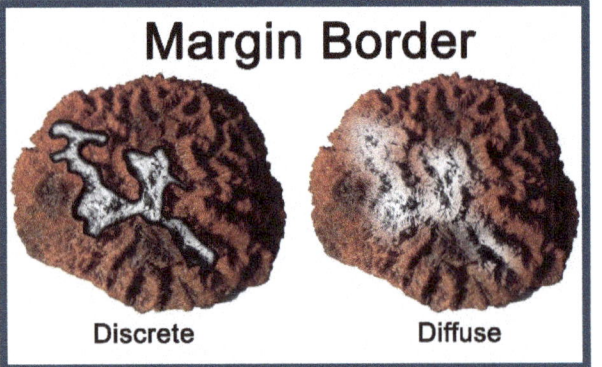

(5) The shape describes the way in which the affliction or banded disease is spreading through the colony, and is described in 3 ways:
- **Linear:** moving in a line across the colony
- **Annular**: Spreading as a circular ring
- **Irregular:** not having any defined shape

(6) The border of the lesion is described as either **Discrete** (obvious, thick, defined) or **Diffuse** (feathered, gradient, difficult to differentiate).

IDENTIFYING PROBLEMS WITH CORAL HEALTH

Corals are a unique organism in that they do not just get old and die. Even though individual polyps may come and go, through asexual reproduction the same colony of coral can live for thousands of years. This means that when a coral colony dies, it is due to some cause. Our job, as researchers, is to try and look at the evidence available to determine what that reason was. In many cases it can be obvious, such as mass bleaching or a tsunami, but in the case of small scale or chronic stresses this can be much more difficult. In these cases you must use all of the available evidence to decide how to code the injured, dead, or diseased corals during reef surveys.

Coral diseases are very difficult to identify, and are often mis-identified even by experienced researchers. In most areas disease is not the leading cause of mortality, so most of the problems you see on the reef will not actually be disease. Often, disease surveys are also known as 'Compromised Coral Health' surveys, in order to also include these non-disease related causes for coral degradation or mortality. Before you can begin to identify disease you should first review some of the other causes for declines in coral health so that you do not confuse them with disease. It is strongly recommended that you also download a free copy of the *Underwater Cards for Assessing Coral Health on Indo-Pacific Reefs*, published by the CRTR and available on **GEFCoral.org**. After each category listed below is the code for recording during the EMP survey.

1. PREDATION (PR)

One of the most regularly observed causes for tissue loss or mortality in corals is due to predation. Predation can either occur quickly as in the case of the Crown of Thorns Starfish, or be a chronic problem over time as with *Drupella* snails and Butterfly fish. Often beginning students have difficulty identifying predation from bleaching, one of the main differences is that in predation there is no tissue over the skeleton, while in bleaching there is, even though it may be hard to see. Look closely to see if the skeleton of the coral is exposed or not. Corals that have been predated upon will often display feeding scars that can be linked to the animal responsible. Once you are able to identify these you will be able to code predation during the EMP. For example, if you found a mushroom coral that was recently killed and you see a Crown of Thorns Starfish nearby you would record that point on the substrate survey **HC R RKC FUNG PR COT** [Hard Coral, Solitary, Recently Killed, Fungia, Predation, Crown of Thorns]. Below is a list of the most common coral predators on Koh Tao, for more information on their feeding behavior see the accompanying slideshow or the *CRTR Coral Health Booklet*:

- ***Drupella* Snails (DRUP)** – Generally identified by predation originating from the base of the coral and moving towards the more exposed areas. Because they have a small feeding apparatus (the radula) there will generally be no feeding scats on the skeleton. Usually the individuals are easy to find, except after large storms.

- **Crown of Thorns Starfish (COT)** – large sections of coral consumed, generally from a limited number of species. Crown of Thorns feed by inverting their stomachs and secreting digestive juices which breakdown the living coral tissue, which are then 'sucked' back up. This means that the COTs do not leave feeding scars in the skeleton.

- **Butterfly fish (BTYF)** – Appears as a chronic progression as the fish start with a compromised polyp and progressively consume adjacent polyps. This type of predation is very easily confused with White Syndromes (discussed below).

- **Triggerfish (TRIG)** – Generally appears as a large section of coral broken off, generally the summits of submassive Porites, with exposed holes where previously clams/worms resided. Often the broken areas are surrounded by exploratory scratches. Pieces of the coral may be found below, as generally the fish are after the invertebrates living in the coral, rather than going after the coral itself.

- **Damselfish (DAMS)** – Appears as chronic predation as the fish kill the coral polyps and then uses the dead coral skeleton to grow filamentous and macro-algae or deposit its eggs.

- **Parrotfish (PRT)** – appears as many oval shaped lesions, generally along the peaks and ridges of a coral where zooxanthallae are more abundant. They may or may not leave feeding scars in the coral skeleton, depending on the size of the fish.

- **Unknown (UKN)** – Any coral that has had the tissue removed but does not fit with any of the patterns/mechanisms above.

2. Overgrowth and competition (OG)

Overgrowth and competition is also a major reason for coral decline. Usually this is easy to identify as the organism overgrowing the coral will still be present. The following are some of the common organisms that will overgrow and smoother or kill corals:

- Macro-Algae (MA)
- Cyanobacteria/Filamentous Mats (CB)
- Tunicates/Ascidians (TA)
- Sponge (SP)

3. Others

- **Physical damage (PHY)** – includes all non-predation types of physical damage to the coral, usually caused by divers, debris, anchors, etc.
- **Sedimentation Damage or Smothering (SDS)** – Silt or sand smothering the coral, usually caused by increased run-off following storms, land development, deforestation, etc. May also be caused by divers or boats going over shallow reef areas.
- Unknown (UNK)

Coral Diseases of the Indo-Pacific

Now that you know how to identify and code the non-disease health problems for corals, it is time to start looking at some of the diseases relevant to our area. Keep in mind that this is a quickly developing field and many of these descriptions may have changed since publication. This is not a complete list, but is an introduction to some of the most prevalent diseases and how to identify them. The rise in observed coral disease outbreaks has been closely correlated with anthropogenic activities, as stressed corals do not have healthy immune systems to protect against disease.

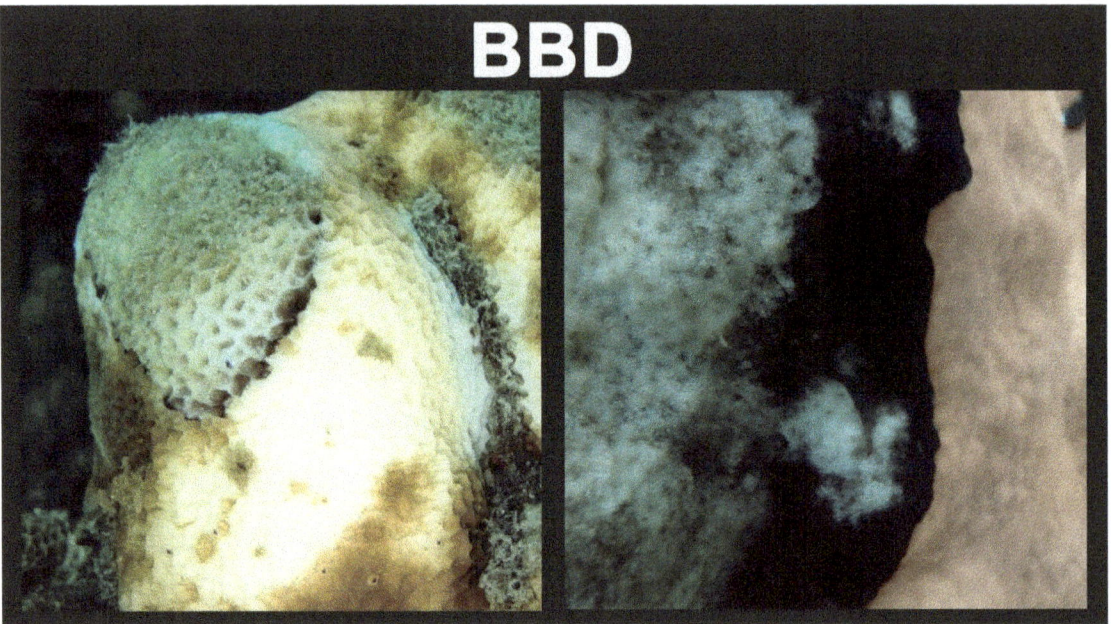

Black Band Disease
Code: BBD

Description: This disease was first observed in the reefs of Belize in 1973, but has since been observed on most major reefs around the world. It is characterized by black or reddish filamentous mats, 1-3 cm in width, annular or linear. The focal infection usually begins in a damaged area on the tops or sides of the colony, and rarely begins in healthy tissue. The band may progress from 3mm-1cm per day. Affects at least 40 species of coral, and is one of the most problematic diseases affecting corals worldwide.

Microbial Agent: Various micro-organisms (known as a consortium); predominately cyanobacteria and also sulfide-oxidizing and sulfate-reducing bacteria.

Abiotic Controls: As Cyanobacteria are one of the more abundant microbes involved in this disease, it is Usually confined to well illuminated corals. Occurs only when temperatures exceed 29°C, occasionally subsiding during annual temperature fluctuations. Increased rates of disease progression have also been linked to nutrient enrichment (particularly sulfur-rich environments).

Other notes: In 1987 a BBD aspirator (underwater suction device) was tested to remove the filamentous mat. After complete removal, modeling clay was placed over the lesion border. It was found that this technique was 70% effective at saving infected corals.

YELLOW BAND DISEASE

CODE: YBD

DESCRIPTION: Appears as a pale yellow blotch, which then radiates outwards, progressing slowly. May be focal or multifocal, with a diffuse border. The disease progresses by killing the zooxanthallae by preventing mitosis. Has reached epidemic proportions in the Caribbean during the 1990's, but only known to affect *Diploastrea Heliopora*, *Fungia spp.* and *Herpolitha spp.* in the Indo-Pacific Region. The disease affects stressed corals.

MICROBIAL AGENT: A consortium of 4 species of Vibrio Bacteria

ABIOTIC CONTROLS: Unknown, but possibly linked to increased water temperatures.

Brown Band Disease
Code: BrB

Description: A brown band consisting of mobile ciliates which occur at the border and over the first few cm of recently killed coral. Generally the disease begins from the center of branching corals and progresses towards the tip. The band and lesions often are diffuse and sometimes there may be a white band between the brown band and living coral tissue. Most commonly observed on Branching *Acropora* corals, but has also been observed on *Pocillopora* and *Favidae*.

Microbial Agent: Ciliates (protozoans with hair-like organelles called cilia) and bacteria. The ciliates consume the coral and also harbor photosynthetic zooxanthallae. Studies of BrB have identified 11 different species of ciliates in the Great Barrier Reef, 4 of which contain zooxanthallae, giving the band its brown color. At this time, it is still unknown which organism (bacteria or ciliates) are the causative agents and which is an advantages agent, but both are found together in both Brown Band and White Syndrome.

Abiotic Controls: Unknown

Skeletal Eroding Band
Code: SEB

Description: Black or dark green spotted band, with a 'peppered' appearance. Close up the band appears very different from the mat like appearance of BBD. The band may be several mm or cm thick, with a strong to light gradient starting from the living coral tissue. Often the band begins at the base of the colony, and in some cases has been associated with white syndromes on the same colony. Most often affects *Acropora* and *Pocillopora*, but has been recorded on over 26 genera and 12 families of coral. First identified in 1988, SEB is now the most commonly recorded disease in the Indo-Pacific. The disease progresses at a rate of about 2mm per day, and has a 95% morality rate.

Microbial Agent: Boring Ciliate (*Halofolliculina corallasia*)

Abiotic Controls: Unknown, although correlated with warm/polluted waters

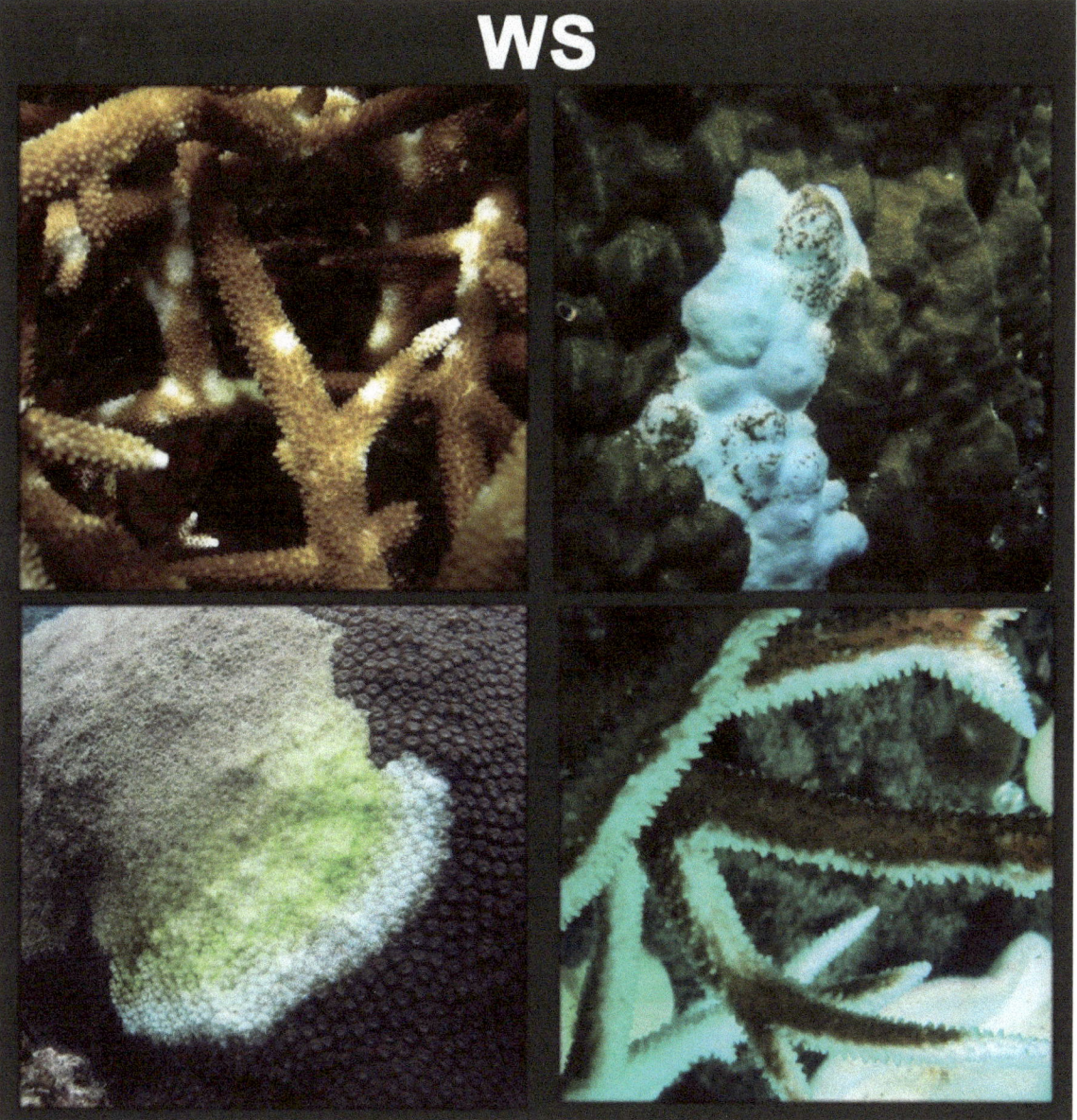

White Syndromes
Code: WS

Description: Diffuse and complex patterns of bleaching and tissue loss occasionally results in coral tissue at the margin sloughing off the skeleton. The pathogens involved in White Syndrome are different from the White Band Disease, which essentially decimated the Acropora of the Caribbean in the 1970's. The term White Syndrome is specific to the Indo-Pacific, and includes White Patch, White Pox, and White Band which used to be thought of as separate diseases. For the last decade, this has been the most prevalent disease throughout the Indo-Pacific.

Can progress up to 2 cm per day in some cases, and is known to infect at least 15 genera of corals throughout the Indo-Pacific. Easily confused with chronic predation by *Drupella* Snails or Butterfly Fish. Currently, is the most prevalent disease on the Great Barrier Reef, affecting over 11% of corals.

MICROBIAL AGENT: Several types of bacteria (including *Vibrio*) and ciliates. Communities are very similar to BrB, except that the cilliats do not contain zooxanthallae.

ABIOTIC CONTROLS: Warmer temperatures or large annual temperature variations.

UNUSUAL BLEACHING PATTERNS
CODE: UBP

DESCRIPTION: Distinguished by focal or multifocal bleaching patterns of unknown causes, when surrounding corals are healthy. Often appears on Porites, but can also be found on other coral. Indicated by the coral tissue being pale or white, usually has a distich border, and occurs when sea water temperature are within a normal range.

MICROBIAL AGENT: Possibly bacterial

ABIOTIC CONTROLS: Unknown

Unexplained Growth Anomalies

CODE: GA

DESCRIPTION: Enlarged, deformed, or abnormally arranged polyps and coral skeletal elements (corallites, ridges, coenostreum, etc.). Tumor-like growths called nodules. Pigmentation may be normal, pale, or brightly colored due to decreased concentrations of zooxanthallae. Causes partial mortality and decreases fitness, growth rates, and fecundity (reproductive output). Recorded in most genera of coral in both the Caribbean and the Indo-Pacific. Acropora and Porites corals seem to be some of the most susceptible to GAs.

MICROBIAL AGENT: Unknown or possibly none

ABIOTIC CONTROLS: Largely unknown, however a comprehensive 2011 study by Aeby *et al*. found that in *Acropora* the prevalence of GAs was mostly correlated with the density of Acropora coverage (GAs were most prevalent in areas with the most Acropora corals.) Not factors such as proximity to human population density. In Porites however, GAs could mostly be explained by proximity to major human population centers (36.8%) followed by density of Porites corals in the area (19.9%) and degree of exposure to UV radiation (16.6%).

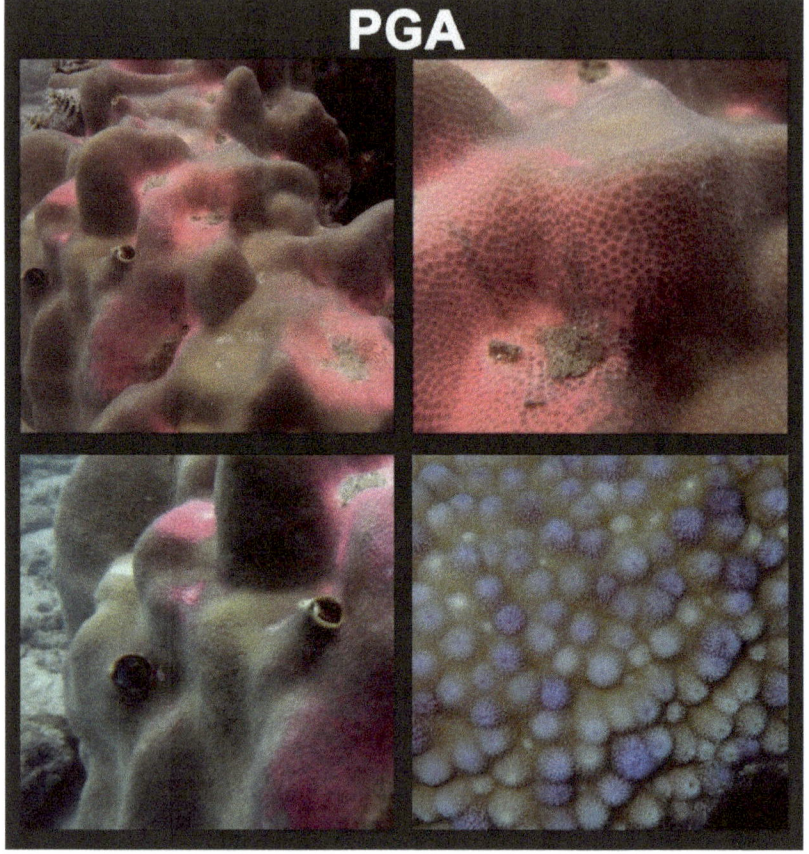

Pigment Abnormalities

Code: PGA

Description: Multifocal or diffuse areas of bright pink, blue, or purple color. Is not a banded disease, but is a general immune response to physical or pathogenic stressors. Like most invertebrates, corals lack a specified immune system, and rely on a generalized approach to defending against invaders. PGAs are not usually fatal, but can persist for years. Usually observed in *Porites* or *Acropora* corals

Microbial Agent: Varies

Abiotic Controls: Varies

Trematodiasis

Code: TMD

Description: Bleaching, combined with small focal pink and white swelling of tissue and formation of nodules. Slow progressing, but can persist for years or decades. Only observed on submassive *Porites*.

Microbial Agent: Parasitic Trematode (flatworm)

Abiotic Controls: Unknown

* * *

Now that you know what coral diseases are and how to identify them, you can begin to record them during the EMP Substrate Surveys. However, in a normal year it is very unlikely that a diseased coral would be lying directly under the transect line. In the case of diseases, we sometimes also modify the techniques used in order to increase the number of corals sampled during the survey. In the next chapter you will be introduced to some of those modified techniques that can be used for specialized surveys such as diseases and compromised coral health.

Chapter 8 Review

After completing the reading and discussion of the material covered in Chapter 8, you should understand and be able to complete the following exercises. Please talk with your instructor about any questions you may have.

1. Look back at the first few pictures in this chapter, are they afflicted by predation or disease? Is it chronic or acute? If it is a diseases, which one is it?

2. For the picture above, please describe the lesion using the terms from this chapter:
    a. Lesion Pattern:
    b. Rate of Progression:
    c. Lesion Color:
    d. Margin Color:
    e. Band Thickness:
    f. Shape:
    g. Border:

3. Find at least one diseased coral on your next survey dive and point it out to your instructor.

Chapter 9: Specialized Reef Survey Methods

Chapter 9: Specialized reef survey methods

Introduction to advanced survey techniques

The EMP Survey methods allow us to run a robust monitoring program which is scientifically valid, yet is easily taught to a wide range of skilled divers. As a program that includes a variety of important indicator species and coral health states, it is generally all that we need to provide the long-term data that is used for research projects, technical reports, and local policy decisions. Sometimes, however, we need to use modified or advanced survey techniques in order to research specific organisms or reef processes. In this chapter we will introduce you to several advanced survey techniques. Once you are familiar with the techniques, you should have the foundation to also develop your own research methods for specific research questions you may wish to investigate.

Photographic Surveys

With underwater cameras becoming less expensive and more user-friendly, they are also becoming an important tool for reef managers and researchers. Photographs can be analyzed in many different ways and also sent to other researchers all over the world. They are so useful, in fact, that cameras are required equipment for any professional reef researcher. Some of the benefits of photographic surveys include:

- Providing a record of the current state of reefs
- Providing a record that can be analyzed by future researchers
- Getting proof of observations and first records
- Being able to look up unknown organisms later
- Increasing the number of points sampled in substrate surveys
- Tracking growth rates, colonization rates, etc.
- Providing graphics that can easily be understood by non-scientists

Taking scientific photographs is different from normal underwater photography in many ways. In scientific photographs there should be consistency with the way that the photographs are taken, and often they must include a scale. In this section you will learn two techniques for taking scientific photographs during your dives, those taken along a transect line and those taken of or from specific points.

Ecological Monitoring Program Manual

1. PHOTOTRANSECTS

Phototransects are photos taken along a transect line, they can be used when researchers want to increase the number of points sampled, or when researchers may be analyzing reef data remotely or at a later date. By providing a record of the transect line, researchers later on may also be able to analyze variables that we don't yet know about, or are not currently interested in. For example, a researcher may want to look at the changes in types of macro-algae growing in an area over the last several years, but our EMP data only records it as NIA. By using the Phototransects we have already taken, the researcher would be able to look back at all the sites and then analyze the historical changes in algal species.

Phototransect images can be analyzed using a variety of programs, one of the most popular being the free software called Coral Point Count (CPCe). In this program, a set number of randomly distributed points are overlaid on top of the picture, and the researcher codes each one just as we would do underwater during

AN EXAMPLE IMAGE FROM A PHOTOTRANSECT SURVEY, SHOWING THE 10 RANDOMLY LAID OUT POINTS IN PHOTO-ANALYSING SOFTWARE SUCH AS CPCE. NOTE THE LOCATION AND OREINTATION OF THE PHOTO TO THE TRANSECT LINE, WITH THE IMAGE CENTERED AT 18M.

the EMP. During the EMP Substrate survey, only 160 points are analyzed, which seems like a lot while you are doing the survey, but means that any organisms not covering more than 1% of the bottom are not likely to be recorded. If a Phototransect was conducted along the same line it would yield 80 photographs (1 photo every meter X 20m X 4 sections) with , and if each was overlaid with 10 points, then a total of 800 points would be analyzed. This greatly increases the chances of recording organisms that cover only a small amount of the coral reef. By surveying from a larger area, the Phototransect also allows for greater statistical precision between surveys, as it is not as highly influenced by where the line is laid out as the standard substrate survey is.

CONDUCTING THE PHOTOTRANSECT SURVEY

The most important part in conducting the Phototransect survey is to have consistency between all of the photographs. This means that each picture is taken parallel to the surface being photographed, as not to create bias. Furthermore, each photograph will need to have the same scale; to achieve that, *we always take the photograph 1 meter from the bottom.* You will also have to set the ratio of the photos to 1:1, the default for most cameras is 4:3 or 3:2, which results in a rectangular image. In some cases the 'Custom White Balance' feature can be used, but generally it is preferred to simply set the camera to 'Underwater' white balancing mode. Your instructor can help you set up and use additional features on your camera before heading out for the dive.

The distance between photographs will vary depending on the specific survey, but if you are taking a Phototransect of an EMP line, we take the photo every 1 meter, beginning at 0 and following the same segment breaks as the normal survey.

To standardize the procedure, we always take the photographs on the right-hand side of the line. Phototransects should be taken at each EMP site 2-3 times per year.

Often, the hardest part of this survey for divers learning it is to keep the camera parallel and 1 meter from the bottom. In some cases it may be helpful to bring a 1 meter PVC tube with you on your first few dives, but be very careful not to damage the reef with it. We recommend practicing in a sandy area first, until you can accurately maintain a distance of 1 m from the sea bed consistently. Some researchers will also use frames to hold the camera and achieve a perfect distance and perspective each time, however we discourage this as it greatly increases the contacts and damage to the reef.

2. Photo-documentation

Photo-documentation can be used to calculate the size of a variable or to track changes over time. If using photos to calculate the size of an organism underwater, the photo must contain at least one reference with known dimensions, usually a ruler. The reference must be in the same plane and perspective as the object being photographed, so that its true size is not distorted in the 2-D image. The image can then be analyzed using the free software *CPCe* or *imageJ*. In both programs, users set the scale of the photograph using the reference (ideally both a horizontal and vertical reference), and the program then calculates the area of the object in mm^2 or cm^2.

To use photo-documentation to track changes over time, the photographs must be taken in the same way and from the same location each time. If you are based at the reef you are monitoring then you may be able to remember where and how the pictures of reef features need to be taken, but generally they will need to be marked. If you plan to use Photo-documentation in a research study, you should include the set-up of the site in your initial plan, and decide where the photos should be taken from to achieve the best

results. The locations where the photographs will be taken from can then be marked with a metal stake driven into the sand, or a stainless steel nail driven into a non-living solid object. Each time you take the photos, set the camera on the marker and ensure that it is pointing in the same direction each time.

Alternatively, the original pictures can be printed and laminated to bring underwater along with a simple map. When you identify one of the corals that have been mapped, refer to the laminated pictures to establish the same shot (perspective and distance).

Quadrant Surveys

Quadrant surveys are generally used to calculate the coverage of some reef benthos, or to aid in the counting of small and cryptic organisms. Essentially, the quadrant creates a border around a section of reef, and divides it into smaller sections that are easier to analyze. A quadrant can be made any size; however a 1 meter quadrant broken into 25 grid sections (5 X 5) is most common for reef monitoring studies. Quadrants can either be used randomly, or installed permanently to track variables over time. Some of the common applications for quadrants include:

EXAMPLE OF HOW TO USE A QUADRANT TO ESTIMATE PERCENTAGE COVER, IN THIS CASE OF THE CORAL *GONIOPORA* (OUTLINED IN ORANGE)

- Calculating coral, algal, or sponge coverage
- Tracking the spread of fouling organisms
- Tracking recruitment and colonization rates
- Counting *Drupella* snails on the reef
- Assessing coral disease or other health factors

To use the quadrant, carefully lay the PVC frame over the reef area and estimate the percent coverage in each grid section as shown in the image at left. In order to estimate accurately, be sure that you are making observations perpendicular to the plane of the frame. The frame can be laid out randomly, haphazardly, or along a transect line. Quadrants can also be used in conjunction with photo surveys, or installed permanently (generally using steel pegs to mark the corners and small rope to designate the borders.)

The percent coverage can be calculated by recording the estimated coverage in each section, and then averaging all the sections. In the example photo shown at left, the average coverage of *Goniopora* is thus calculated as 21%.

When using the quadrant to count small or cryptic organisms, the number in each section is counted, and then the totals summed to get the abundance per square meter.

Benthic Surveys in Non-Reef Areas

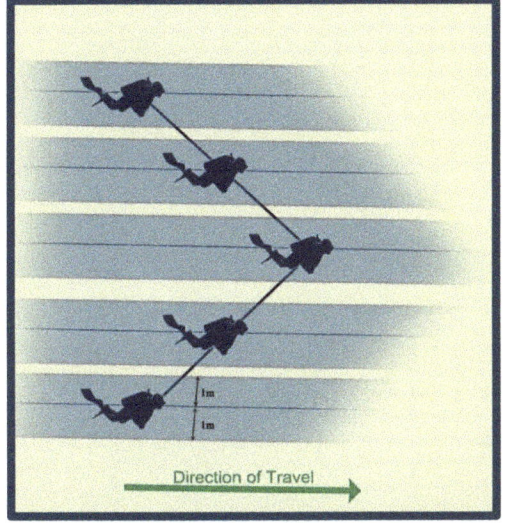

Benthic surveys in non-reef areas (sand, muck, etc.) are used when assessing cryptic, hard to find, or rare species. The goal of the survey is to cover as much area as possible, while still being able to quantify the area observed. This may be used in situations such as assessing the populations and diversity of nudibranchs or seahorses. In this survey, each diver observes data within an imaginary belt, also known as a run. In the example at left, each run is 2 meters wide, as the observers look 1 meter to either side of their body. The observers are positioned along a guide line so that they all cover the same amount of area, and so that their runs do not overlap. The divers are arranged in a 'V-Formation' rather than a straight line so that it is easier for the leader (the person in the center) to control the entire team's position and movement.

In addition to knowing the width of the diver's runs, the length of the survey will also need to be known in order to calculate the total area covered. This can be accomplished in several ways:

- Towing a GPS on a raft at the surface
- Beginning and starting the survey at known points, or taking a GPS position of the starting and ending points
- Timing the survey and factoring together with the average distance per traveled per minute
- Counting kick cycles

When it is impractical to calculate by distance, time can also be used to calculate observer effort. When using timed surveys, try to plan a preset time for each survey/location.

Coral size-class, recruitment, and fragment survey

[Note: this section co-authored by Florian Lang, 2014]

Survey description

The substrate survey of the EMP is done along a point-intercept transect, which gives information on the percentage of coral coverage, but does not take data on the size or life stages of the corals. This is instead done in another survey along 1x20m belt-transect on the normal EMP transect lines, known as the coral size-class, recruitment, and fragment survey (SRF Survey). This survey should be done about one time per year for each EMP site.

Coral size class is a measurement of the total colony size of each coral in the belt survey, expressed as the maximum colony diameter in centimeters.

The class size is important because many of the life-history processes of the coral are dependent upon or related to colony size, such as reproduction. Furthermore, the size-frequency of the corals provide information on the history of the reef and the reef's resilience to current or historic threats and disturbances. On a healthy and functioning coral reef, we would expect to see uniform distribution of class-size frequency – the reef having corals of all sizes. In degraded reefs we may see that there is a higher proportion of larger corals, as the small ones would have succumbed to threats. In a reef recovering from major mortality we may find only small and medium sized corals. Any changes in the size-frequency of corals at a site should be investigated further.

Recruitment refers to the number of successfully settled coral larvae on the reef. Reproduction and recruitment is the most important factor in determining the long-term trajectory of reef health and biodiversity. It is through the successful recruitment of coral larvae that reefs are able rebound from disturbances, maintain biodiversity, and adapt to changing environmental conditions. Recruitment failures commonly result when adult coral populations become too sparse and isolated, or there is a lack of available substrate due to overgrowth by macro-algae. For the recruitment survey, all corals less than 10 cm which do not appear to have suffered partial mortality or be produced through fragmentation are counted as recruits.

Fragmentation is the asexual reproduction process whereby sections of a coral colony are broken off, but survive and begin to grow as a separate, but genetically identical, colony. For some coral genera such as *Acropora*, *Pocillopora*, and *Pavona*, fragmentation is a major strategy to increase the coverage of the colony. In fact, branching *Acropora* corals will often form what are known as mono-specific stands, or clusters of branching colonies that are all genetically identical and produced through fragmentation. However, a high abundance of fragments on a reef, or fragments of massive and submassive corals, may indicate natural or human related structural disturbances.

CONDUCTING THE SRF SURVEY

The SRF survey is conducted at the same time as the EMP surveys, but by a separate dedicated buddy team. In addition to the usual EMP materials, each member of the team should also have a 1 meter long PVC tube, ruler, and a slate specially set-up for this survey. One diver will be responsible for taking the size-class, and the other for the recruitment and fragment survey. Both surveys are conducted in the same sections as the normal EMP, using the 1m PVC tube centered on the transect line to create the 1m wide belt. All coral colonies, fragments, or recruits in which more than half the colony falls within the belt are recorded, as shown in the image at left (green corals would be recorded and red ones excluded).

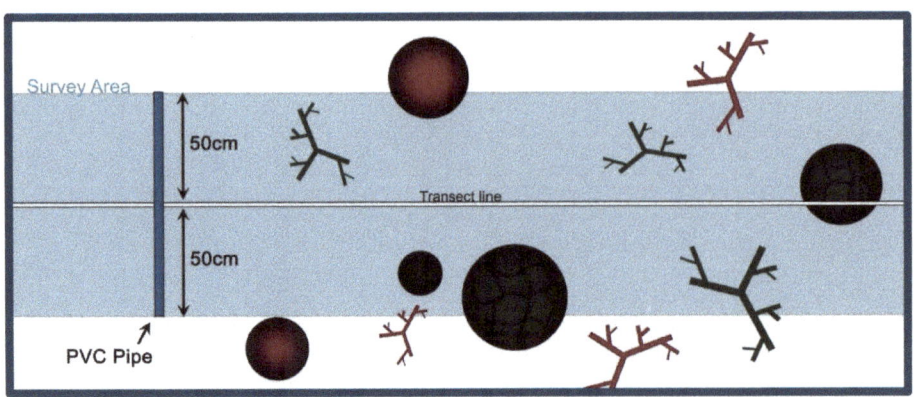

1. Size-class

For the size-class survey, you will need to mark the PVC tube (using electrical tape or black marker) to mark off 10cm, 20cm, 30cm, and 50cm to create the following size-class categories:

- **1**= 10-20 cm
- **2**= 21-30 cm
- **3**= 31-50 cm
- **4**= 51-100 cm
- **5**= >100 cm

(Note; corals under 10cm will be assessed in the recruitment or fragment surveys)

For every established coral colony falling within the survey area, measure the maximum colony diameter using the PVC tube and add a tally to the appropriate column on the slate (shown at left). Mushroom corals are not included in this survey, nor are the large stands of *Pavona* coral which are difficult to differentiate from one colony to another.

Coral Colony Size-class

	Colony Max Diameter				
	1 10-20cm	2 21-30cm	3 31-50cm	4 51-100cm	5 >100cm
ACRO					
ASTREO					
DIPLO					
ECHINPO					
EUPHY					
FAV					
FAVIT					
GALA					
GONIA					
GONIO					
HYDNO					
LEPTO					
LEPTS					
LOBO					
MERU					
MONTA					
MONTI					
PACHY					
PAV					
PECT					
PLATY					
POCI					
POR					
TURB					
UKN					
OTH					

2. Recruitment

For the recruitment survey, use the same 1 meter PVC tube to create the 1 meter belt and use a ruler to measure all recruits. Recruits are settled coral polyps or colonies that are less than 10 cm in size (maximum diameter). You do not need to measure the exact size of the recruits, but instead measure so that you can tally them in the appropriate location on the slate according to coral genera and the following size-class compartments:

- **1**= 0 - 2 cm
- **2**= 2.1 - 4 cm
- **3**= 4.1 - 6 cm
- **4**= 6.1 - 8 cm
- **5**= 8.1 - 10 cm

Recruitment and Fragment Survey

	Recruits					Fragments
	1 0-2 cm	2 2.1-4cm	3 4.1-6cm	4 6.1-8cm	5 8.1-10cm	
ACRO						
ASTREO						
DIPLO						
ECHINPO						
EUPHY						
FAV						
FAVIT						
FUNG						
GALA						
GONIA						
GONIO						
HYDNO						
LEPTO						
LEPTS						
LOBO						
MERU						
MONTA						
MONTI						
PACHY						
PAV						
PECT						
PLATY						
POCI						
POR						
TURB						
UKN						
OTH						

3. Fragments

For the fragment survey, count any corals that have been broken off of a larger colony and record them in the appropriate location on the slate, you do not need to measure the fragment sizes.

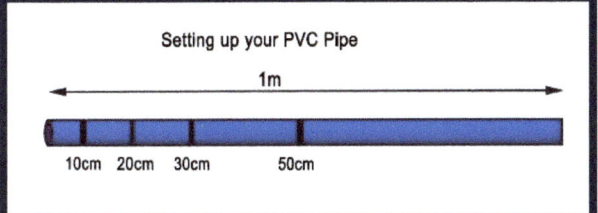

Setting up your PVC Pipe — 1m — 10cm 20cm 30cm 50cm

THE COMPROMISED CORAL HEALTH SURVEY (DISEASE SURVEY)

The Compromised Coral Health/Disease survey is performed to get a more in-depth understanding of coral health than is usually obtained with the normal EMP. Generally, it can be performed at each site once or twice a year, or when warranted by EMP data or observations that show a high amount of unhealthy corals in an area. Often this can be caused by problems originating from land (deforestation, development, run-off, etc.) or when there is a suspected disease or predator outbreak.

The survey is conducted along the permanent EMP lines or lines laid down haphazardly in areas were problems are being quantified. If incidence rate or mortality rate are to be calculated you must mark the ends of the transect line so that it may be laid out in the same position at a later date. For this survey, you will use the same method to establish a 1 m wide belt as in the SRF survey as described above. However, since you will be evaluating every coral colony of which the majority is contained within the belt, this can be quite a lot of data to record. For this reason, usually you will only survey the first 10m of each transect section (except in sandy areas where corals are scarce, in which case you should extend the distance to obtain enough corals in your sample). Also, you will modify your slate to allow you to move faster. Instead of having to code each coral (HC S PBL POR OG TA) which can be quite tedious and time consuming, you slate is set up as a chart with the coral genera on one side and the possible problems on the other (see Appendix E).

Begin at the 0m mark on the line with your 1 m PVC pole centered on the transect line to allow you to visualize the 1 m belt. Move forward until a coral falls within your belt (the center of the coral is within range of your stick). Next, assess the entire coral for signs or disease or compromised health (damage, overgrowth, smothering, etc.) record this data with a check mark on your slate, note that sometimes the coral may have more than one problem affecting it and you can only mark it in one box, in such cases choose the category which seems to be the primary problem. Don't forget to also mark the healthy corals so that at the end of the survey you can calculate the prevalence (# of diseased coral/total # of corals assessed). Occasionally on some reefs you may find many of the corals have the same two problems (i.e. sedimentation with pigment anomalies. In such cases, you may also add a row to your slate for that (e.g. POR SDS/PGA).

Continue assessing corals until you reach the 10 meter mark on the line. In some cases you may need to shorten (lots of coral colonies) or extend (mostly sandy areas) the distance you travel along the transect line so that you finish the 4 surveys at your pre-planned air level and required minimum number of colonies assessed. However, when doing so, be sure to record the distance that you surveyed on the line so that you will able to calculate the number of diseased corals per m^2 at the end of the survey. Diseased or compromised corals can also be marked, using stainless steel nails or epoxy, and photographed to assess disease progression. To do so, hammer in a nail at the edge of the dead coral area and record the date. At later dates of observation measure the distance from the nail to the disease border.

MASS CORAL BLEACHING SURVEYS

For the EMP you learned how to identify coral bleaching as Partially Bleached (PBL) or Fully Bleached (FBL). In general, during the EMP you are not going to find many bleaching corals, and so this is sufficient. However, years with thermal anomalies are occurring more frequently, and seem to be becoming more severe, as evidenced by the global bleaching of 1998, 2010, and 2015-2016. In such years, we want to record the bleaching event in more detail so as to contribute to the global effort to document and better understand bleaching events. Ideally, these surveys should be performed before the bleaching event (when sea water temperatures are above normal), during the bleaching event, and after the bleaching event (until all bleached corals have either recovered or died).

Bleaching surveys can be done on the permanent transect lines, or done haphazardly to document areas of significant bleaching, or areas which do not seem to bleach as extremely. The survey can either be done as a point survey, like the EMP, or the belt survey. Which survey you decide to do depends on your preferences, and the abundance of corals in the area. In areas where corals are less abundant, the belt survey will yield more samples. Like the Compromised Health/Disease Survey above, we will modify the slate to speed up the survey and make the work less tedious underwater. An example slate set-up can be found in Appendix F.

When performing the bleaching survey, you will assess each coral under the line/belt and make a tally mark in the appropriate box as follows.

- **Healthy** – Code for any corals which still retain their normal color.
- **Partially Bleached, Living** – Code for corals which still have intact tissue, however have lost pigmentation due to the loss of symbiotic algae. Unlike the EMP, you would also record the amount of the coral (as a percent) and the type of pigment loss:

  - **Pale** - The coral is not white, but has decreased pigmentation. Record the amount of the coral which is pale as a percentage (ie if the top half of the coral is pale then record as 50%).
  - **Molten** – Molten refers to white patches or spots on the coral, they are generally multi-focal and have diffuse borders. Record the approximate percentage of the coral which is affected with these spots/patches.
  - **White** – Refers to focal bleaching in which the coral has lost all zooxanthallae in about 1-90% of the colony, often with a diffuse border. Frequently this can be observed towards the beginning or end of the bleaching event with massive or submassive corals due to the interaction of temperature and light on the coral's symbiotic algae. Record the percentage of the colony which is 'white.'

- **Fully Bleached, Living** – As in the EMP, use this code for any coral which no longer contains any Zooxanthallae. You do not need to quantify percentage as it is already assumed to be 100%.

- **Recently Killed/Dead Coral** – As the bleaching event progresses, corals will begin to suffer mortality. In addition to coding the mortality as in the EMP, also code the percent of the coral which is RKC or Dead. Remember that this should be coral which has died during the bleaching event and not that which is older mortality.

Note, it is also possible to have mixed bleaching and mortality, such has half white and half pale (P – 50%, W – 50%) or mostly FBL and partially Recently Killed (FBL- 80%, RKC – 20%).

When performing the Bleaching Survey, it is very important to record as much about the abiotic conditions as possible, including temperature, weather, water visibility, wave/current conditions, and depth the survey was performed. Try to record these variables as accurately as possible. If the program you are at has the capacity to record other water quality variables such as salinity, turbidity, nutrient load, etc., then these should also be accurately recorded. Also, take time immediately after the dive to record your notes and general observations. This may include:

- Any coral genera which were not bleached, or alternatively any that looked less healthy than the general population.
- The occurrence of predation, overgrowth, or other secondary problems for the reef.
- Any other causes of coral mortality such as sedimentation, which has a severe effect on the bleached corals.
- Reasons why the amount of bleaching may vary at this site (i.e. currents, shading, depth, diversity of the coral community, etc.).

- Any signs of recovery.
- Additional organisms which are found to be bleaching (soft coral, anemones, clams, sponges, etc.).

Another great tool is the Coral Watch Health Slate, developed by the University of Queensland and available for purchase at **coralwatch.org.** The slates contain 4 color and 6 intensity codes for the majority of reef building corals in the Indo-Pacific. To use the cards, align the card to the closet color match, and then identify which is the lightest color intensity observed on the coral colony. Next, use the same color to identify the darkest intensity observed on the coral colony. Both values are then recorded on the slate.

Advanced Fish Surveys

For most monitoring programs, the fish survey as described earlier is sufficient to track the biodiversity and abundance of some of the most important species on your reefs. The list of indicator species listed in this manual is a balance between obtaining useful data, while not being too difficult to learn. However, in some cases, such as scientific research projects it can be desirable to collect data from a wider range of species or genera, as they can also be ecologically important. Once you learn all the basic indicator species for your area, it is recommended to start learning about species which were not included, or to break down the genera level groups into species level.

Of course, it would not be practical to try to learn or survey all the fish on the reef, but below we have provided a few families of particular importance. You can use the internet or local fish guides to identify and learn some of the species from the families listed below, and your instructor may provide other lists for the particular area or research study being conducted.

1. Family: *Pomacentridae*

A. **Genus:** *Pomacentrus*- The majority of Damselfishes fall in this group. They tend to be herbivores, and farm algae on dead coral of the reef. A few species may be either planktivorous or omnivorous. The herbivores species sometimes have a negative effect on coral, as they kill coral tissue in order to open space for algae to grow, or to deposit their eggs. They are threatened by over-collection for the aquarium industry.

B. **Subfamily: Amphiprioninae** – also known as the anemone fish, these colorful and ornate fishes have an obligate symbiotic relationship with anemones. They make up about 43% of the ornamental aquarium trade, with only 25% of that coming from breeding programs.

2. Family: *Pomacanthidae*

The family Pomacanthidae is also known as the Angelfishes, they are bright, ornate, and conspicuous reef fish much like the butterfly fish. However, they are bigger and have a more varied diet. Most are omnivores, scraping algae from the rocks and reef surfaces, while also consuming small invertebrates. The juveniles and adults have quite different appearances, but both are quite visually striking and thus important to the tourism industry. They are threatened primarily for live trade in the aquarium industry.

3. Family: *Tetraodontidae*

This group is also known as the pufferfish. Although slow moving, these fish are able to expand their stomachs to deter predators, and many species contain powerful toxins. They are omnivores, feeding primarily on algae, but will also feed on invertebrates, including clams and shellfish. They are closely related to the porcupine fishes.

4. Family: *Diodontidae*

The porcupine fish are closely related to the pufferfishes. They have the ability to inflate their bodies to increase size and avoid predation. They also have spines covering their bodies which protrude outwards when inflated. Some species contain toxins. They are a commercial fish, which are eaten as a delicacy or turned into fish food.

5. Family: *Ephippidae*

This group includes the Batfishes, which are a vital fish to coral reef health. These herbivorous fishes eat macro-algae which is not palatable for parrot or rabbit fishes, preventing it from overgrowing the reefs. The juveniles are morphologically distinct from the adults, and tend to seek cover under debris at the surface. They are a commercially fished species, and also one widely traded in the aquarium industry.

6. Family: Scorpaenidae

This group includes the scorpion and lionfish, and some of the most poisonous fish in the ocean. Both groups are effective carnivores, feeding on invertebrates and small fish. Lionfish have gained a bad reputation in the Atlantic and Caribbean as an invasive predator which has threatened local fish populations. However, in the Indo-Pacific they are not problematic, and are an important part of the reef trophic structure. They are threatened primarily by the aquarium trade.

7. Family: Mullidae

Goatfishes are easily identified by their 'whisker' like barbels that are used to dig into soft substrate for prey. They are usually accompanied by wrasses or other fish which take advantage of the goatfishes digging ability and grab benthic invertebrates as they appear. In areas with a high abundance of mud or silt these fish can greatly increase the water turbidity through their activities.

8. Family: Carangidae

This group contains the Jacks, Trevally, and Pompanos. These are pelagic fishes that inhabit the open oceans but come into the reefs to hunt and visit cleaning stations. The fish are fast, streamlined, and effective predators. They are threatened by over-fishing and consumption.

9. Family: Sphyraenidae

This group, known as Barracudas, are fast, ferocious predatory fish that are widely dispersed throughout the planet's oceans. They are long, slender, stream-lined, and fast. Some species grow to more than 165 cm, and compete with sharks or dolphins for food. They primarily eat fish, from small juveniles to prey as large as the barracuda itself. They are generally pelagic as adults, but may live in large schools on the reef in their juvenile stages. They are threatened by the fishing industry.

Although this is far from a comprehensive list, it should get you started on identifying further important fish species on your reefs. You may have to modify your species list depending on your particular area or interests, and diving the area plus doing your research after dives will help you to create the perfect EMP program for your area.

Chapter 10: Setting up a Permanent Transect Line

Chapter 10: Setting up a permanent transect line

Now that you understand how to conduct the EMP, what species to look for, and how to plan the dives, it is time to learn where the transect locations are. This is a vital skill for performing the EMP since the transects must be laid in the exact same locations and position every month. On your training dive your instructor most likely navigated to the location and showed you where the concrete marker buoy is for the shallow or deep starting points. In the future, it may be up to you to find these locations on your own, or to add new locations.

Finding the transect starting points (point A) on existing lines

As you have seen on your training dives, there are a lot of tasks to be accomplished during the EMP dive, and often students run out of air before completing the entire survey. One of the most difficult tasks in surveys can be finding the Deep and Shallow starting points. If time is wasted while finding the starting locations it could mean that the survey will be incomplete and the data unusable, making for a wasted day of research diving. Sometimes, such as when the viz is not good, this can be a very difficult task, and only a diver with a lot of experience at that particular site will be able to find the points. However, there are three main techniques used to locate the starting points, which are described in more detail below:

- Using provided maps and GPS coordinates (while on the surface)
- Using topography and above surface features
- Natural underwater features and depth profiles
- Search patterns

Using maps and GPS coordinates

Maps and GPS coordinates are the first step in identifying where to begin your search for the transect line starting point. Maps are generally with your instructor, however if you start making your own maps you will find it a useful exercise. Maps will help you to moor the boat in a location close to the survey area, and reduce the amount of time spent finding the site. You should check the maps prior to beginning the dive, and use them when planning the dive or conducting the dive briefing. Each diver should record on their slate the depth of the transect lines and the bearing needed to navigate each line (A-B) and the bearing to move from the Deep line to the Shallow line.

On your map, you will also want supplementary information to help you, including GPS points, natural features or other tips, and any special notes about the site. For example, you may describe the area around the starting point, including any obvious features such as walls, large coral heads, rocks protruding the surface, etc. It is useful to include the distance and bearings from these features to the stating point, i.e. 9m @ 135 deg. Ideally, this should not be a static document, but one that you add to each time something significant happens at the site.

USING TOPOGRAPHY AND ABOVE SURFACE FEATURES

One of the most time-saving methods is to use surface features or features on land to locate the transect starting points. Some of the transects are located close by permanent mooring buoys or can be identified by a rock out-cropping. Often, research divers use triangulation to locate the descent locations to start the survey. If you begin your dive by aligning with these three known points then you will have much less swimming underwater to find the concrete buoy. Each instructor for the EMP will most likely have their own land features they can remember and use, and your instructor will include these in the dive briefing. As a training researcher, it is a good idea to find the point you can use and record those in your dive log book. If you set-up a new transect line in another location this will be a vital skill to master so you can relocate your survey site.

NATURAL FEATURES AND DEPTH PROFILES

Most recreational divers are used to being led around by a dive leader, and never really understand or are able to visualize the site layouts. To test this, just stop halfway through one of your dives and see if you can retrace your steps, or when you get back on the boat and are looking at the surface try to make a mental map of your dive. It is pretty difficult even for those who never get lost on land. As a research diver, after diving a site just a few times you will have a good impression of the features of the site and be able to visualize the site layout. This is because you are taking a more proactive role and concentrating more on the reef features.

During the dive briefing, your instructor will tell you the natural features they use to locate the starting points. As you dive remember to find your own natural features to use in the future as references, after doing this consciously a few times it will become intrinsic and natural for you while diving. One great way to get started is to try to re-locate some of your favorite corals on each dive. Record these in your dive log book and see how many you can find over the next few dives.

Search patterns

U-SEARCH PATTERN

The last resort for finding the transect locations is to use search patterns. Since you will have the approximate depth of the transect line written on your slate, you can then use the search patterns learned in your adventure diving courses to find the permanent starting points. This will of course take time and reduce your air supply, so try to have as much information about the site as possible before beginning.

Setting up a new line

Eventually you may be involved in setting up new transect lines. This may be to build upon an existing EMP Program, starting up your own long-term monitoring program, tracking a disturbance, or possibly the regrowth of a reef after a major problem. In these cases, you will want to find a small research area that is representative of the larger reef area, one that is easy to find, and also safe for divers. In the next sections we will discuss this in more detail, but it helps if you have visited several other EMP sites before beginning.

Finding a suitable location

There are many considerations that should be taken into account regarding setting up a new site for permanent research. First, you need to know what questions your research is set out to answer. If your site is going to be used to track the health of a reef for comparison to other areas in your region then you want them to all be established in similar reefs areas. This means that you would identify different sites, all of which are similar in general features. You would not want to set up several sites in the reef and then have one that is set up in the sand, or at a significantly different depth. In general, sites such as this should be done at a variety of similar locations around your region. If it is an island, then try to find similar areas on all aspects (i.e. North, East, South, and West) so you can get a good idea of how conditions are around the whole island, and to allow for year round monitoring.

In areas where you want to track anthropogenic influences, then you would want to include a few sites near to areas with high development and human activity, and some in more pristine areas. This is why it is important to first familiarize your self with the area, dive or snorkel in as many areas as you can, and also pay attention to surface activities (boat traffic, snorkeling and other types of tourism).

If your efforts are to track a certain disturbance or the subsequent regrowth, you will want to survey the area thoroughly and ensure that you are including the disturbed area, and also an area directly adjacent which is representative of the reef state before the disturbance.

Regardless of your reasons for setting up a permanent monitoring site there are some general rules for doing so. Firstly, you want to find an area that is representative of the entire area. This means that you try as much as possible to not let your own perceptions influence where you set the line, so don't try to choose the 'best' or 'worst' sections of the reef. Generally, we refer to this as 'Haphazard,' meaning it is not random, but also not chosen with any bias. Second, you want to ensure that the area you have chosen is easy to find again. In practice, this means starting near some obvious land features such as moorings, rock outcroppings, etc. Third, be sure that the area is safe for divers, you do not want the transect lines set too deep that divers will have to worry about dive time limits, and you also don't want them too shallow that divers are at risk for boat traffic. You may also want to take other factors into consideration for your specific area, which may include seasonality, distance from dive centers, bottom topography, etc.

CONSTRUCTING PERMANENT A & B POINTS

The starting and ending points of the transect lines need to made permanent, yet not too obvious or obtrusive that divers visiting the area would be upset. It is not recommended to use natural or living objects such as coral heads, giant clams, etc. Ideally, you should create an artificial starting and ending point using concrete and a fishing buoy. Remember that the sea can be extremely rough during

times of the year, turning over large coral heads and moving vast quantities of sand. For this reason, the base should be heavy and partially submerged in the sand. To create the base, use a large washing basin or similar plastic tub, and fill entirely with concrete mixed with plenty of rocks. Concrete loses about 40% of its weight in water, while the rocks do not. Use a piece of PVC pipe to create a hole through the center of the block which will be used to tie the rope through. It is not recommended to put the rope into the concrete while wet, as if it breaks in the future there is no way to replace it. When the concrete is dry, flip over the basin to remove the plastic mold, and then use about 2m of strong marine rope to attach a colorful buoy.

After you have established the location where you plan to deploy the buoys, first send divers down to temporarily mark them with SMBs or floating bottles, so that they can be seen from the surface. Using a kayak or other small craft, drive the buoys to the marked locations, and have experienced divers use lift bags or ropes to lower them to the sea bed, ensuring not to harm the reef. After all 4 points have been

Ecological Monitoring Program Manual

deployed check their positions are correct, then record the bearings and GPS points from a vessel on the surface.

MAPPING YOUR NEW SITE

It is highly recommended for instructors and divers that will continue conducting the EMP after training to have their own journal for each site where they can record notes and make their own personal maps of the site. Things to add to your journal could include:

- Site names, locations, and GPS coordinates
- Land features used to find starting points (and bearings for triangulation if used)
- Condition of the start and end points
- A hand drawn map, including:
    - Depth profiles
    - Reef types encountered
    - Natural features used to locate survey area
    - Favorite corals or interesting reef features
    - Problems, diseased corals found, etc.

EXAMPLE SITE MAPPING AND DESCRIPTION

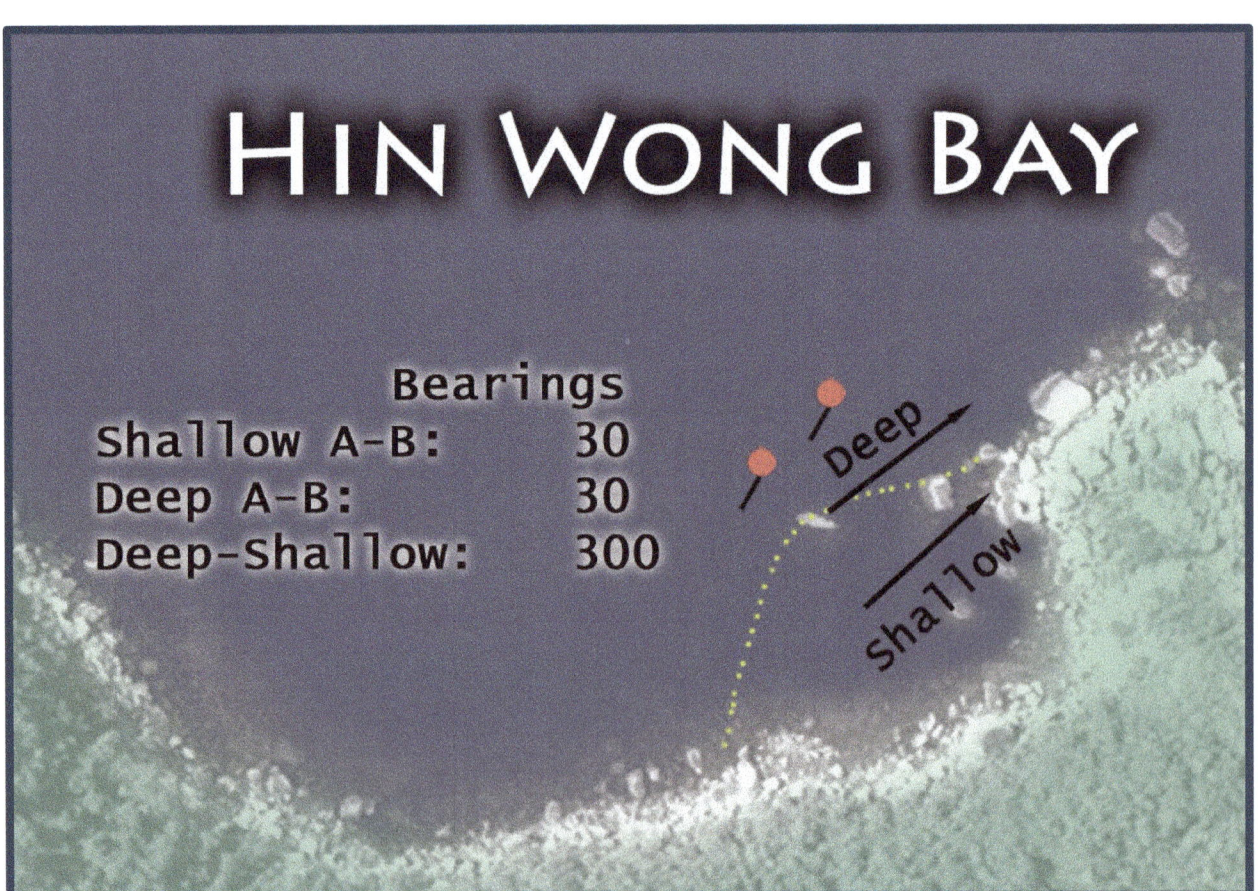

SITE DESCRIPTION

Hin Wong is a secluded bay, with no development in the region of the survey site. The bay does however see a high amount of visitors, primarily snorkeling trips.

The deep reef is dominated by very large massive, submassive, tabulate, and encrusting corals scattered in sand. The shallow reef is dominated by branching corals which, prior to the bleaching event of 2010, had nearly 100 percent coverage of the substrate in many areas.

Transect line locations

Deep Line (aprx. 9-10m depth)

The deep line marker buoy can easily be located, as it is just East of the small rock island where the yellow "no-boat" surface buoy line is attached. The line runs at a bearing of 30° through the patch reef with the 50m point tied close to the boat mooring line. The line then runs over very large massive and submassive corals and ends at a concrete marker at about 9 meters depth.

Shallow Line (Aprx 2-4 m. depth)

From the Deep A point, head up the gradual reef slope at a bearing of about 300 degrees until locating the Shallow A marker buoy in a patch of small, delicate branching corals (*Porites*). From the surface, the marker buoy is located 120 degrees from the face of the very large overhanging rock on shore.

The Shallow line runs at a constant depth through the branching corals, occasionally going over submassive corals. The 50 meter point lies in a group of three submassive corals which form a semi-circle. The last segment of the line runs through a conspicuous rubble patch before ascending up the reef slope just to the right of the large granite boulder. The marker buoy can be found in the canyon created by the large granite rocks at a depth of 2-3m (shown at left).

Ecological Monitoring Program Manual

Appendix and Glossary

Appendix A

Fish and Invertebrate Example Slate Set-up

The Ecological Monitoring Program

Surveyor Name:_____ Survey Location:_____

Survey Date (MM/DD/YY):_____ Survey Time (24hr):_____

Shallow/Deep_____ Locate 'B' Bouy(Y/N):_____

FISH SURVEY REPORT SHEET

		0-20m		25-45m		50-70m		75-95m	
Rays	Note species								
Butterfly fishes	Lined								
	8 - Banded								
	Weibel's								
	Copper Banded								
	Longfin Bannerfish								
Moray Eels	General								
Groupers	Large (> 30cm)								
	Small (< 30cm)								
Parrot fishes	Large (> 20cm)								
	Small (< 20cm)								
Rabbit fishes	General								
Snappers	General								
Surgeonfishes	General								
Sweetlips	General (Juvienelle/Adult)								
Triggerfishes	General								
Wrasse	Red Breasted Wrasse								
	BlueStreak Cleaner								
Other Vertebrates	Sharks, Turtles, Sea snakes								

INVERTEBRATE SURVEY REPORT SHEET

		0-20m		25-45m		50-70m		75-95m	
Giant Clams	Boring								
	Giant								
Echinoderms	Crown of Thorns Starfish								
	Cusion Star								
	Long Spine Black Urchin								
Sea Cucumber	Marbled								
	Black								
	Orange Spiked								
	Pinkfish								
Gastropods	Auger Snail								
	Ramose Murex								
	Nudibranchs/Sea Slugs								
	Drupella Estimate: 0, 1 (1-50), 2 (50-150), 3 (>150)								
Flatworms	Flatworms								
Crustaceans	Hermit Crabs								
Other	Octopus, Cuttlefish, Squid, etc								

Survey Notes: _____

_____

_____

APPENDIX B
SUBSTRATES EXAMPLE SLATE SET-UP

SUBSTRATE SURVEY REPORT SHEET

Surveyor Name:_____ Survey Location:_____ Survey Date (MM/DD/YY):_____ Survey Time (24hr):_____
Transect (Shallow/Deep):_____ Viz: _____ Locate B Bouy (Y/N):_____

Segment 1		Segment 2		Segment 3		Segment 4	
0.5		25.5		50.5		75.5	
1.0		26.0		51.0		76.0	
1.5		26.5		51.5		76.5	
2.0		27.0		52.0		77.0	
2.5		27.5		52.5		77.5	
3.0		28.0		53.0		78.0	
3.5		28.5		53.5		78.5	
4.0		29.0		54.0		79.0	
4.5		29.5		54.5		79.5	
5.0		30.0		55.0		80.0	
5.5		30.5		55.5		80.5	
6.0		31.0		56.0		81.0	
6.5		31.5		56.5		81.5	
7.0		32.0		57.0		82.0	
7.5		32.5		57.5		82.5	
8.0		33.0		58.0		83.0	
8.5		33.5		58.5		83.5	
9.0		34.0		59.0		84.0	
9.5		34.5		59.5		84.5	
10.0		35.0		60.0		85.0	
10.5		35.5		60.5		85.5	
11.0		36.0		61.0		86.0	
11.5		36.5		61.5		86.5	
12.0		37.0		62.0		87.0	
12.5		37.5		62.5		87.5	
13.0		38.0		63.0		88.0	
13.5		38.5		63.5		88.5	
14.0		39.0		64.0		89.0	
14.5		39.5		64.5		89.5	
15.0		40.0		65.0		90.0	
15.5		40.5		65.5		90.5	
16.0		41.0		66.0		91.0	
16.5		41.5		66.5		91.5	
17.0		42.0		67.0		92.0	
17.5		42.5		67.5		92.5	
18.0		43.0		68.0		93.0	
18.5		43.5		68.5		93.5	
19.0		44.0		69.0		94.0	
19.5		44.5		69.5		94.5	
20.0		45.0		70.0		95.0	

Appendix C
Substrate Cheat Sheet

(can be copied and laminated for underwater use)

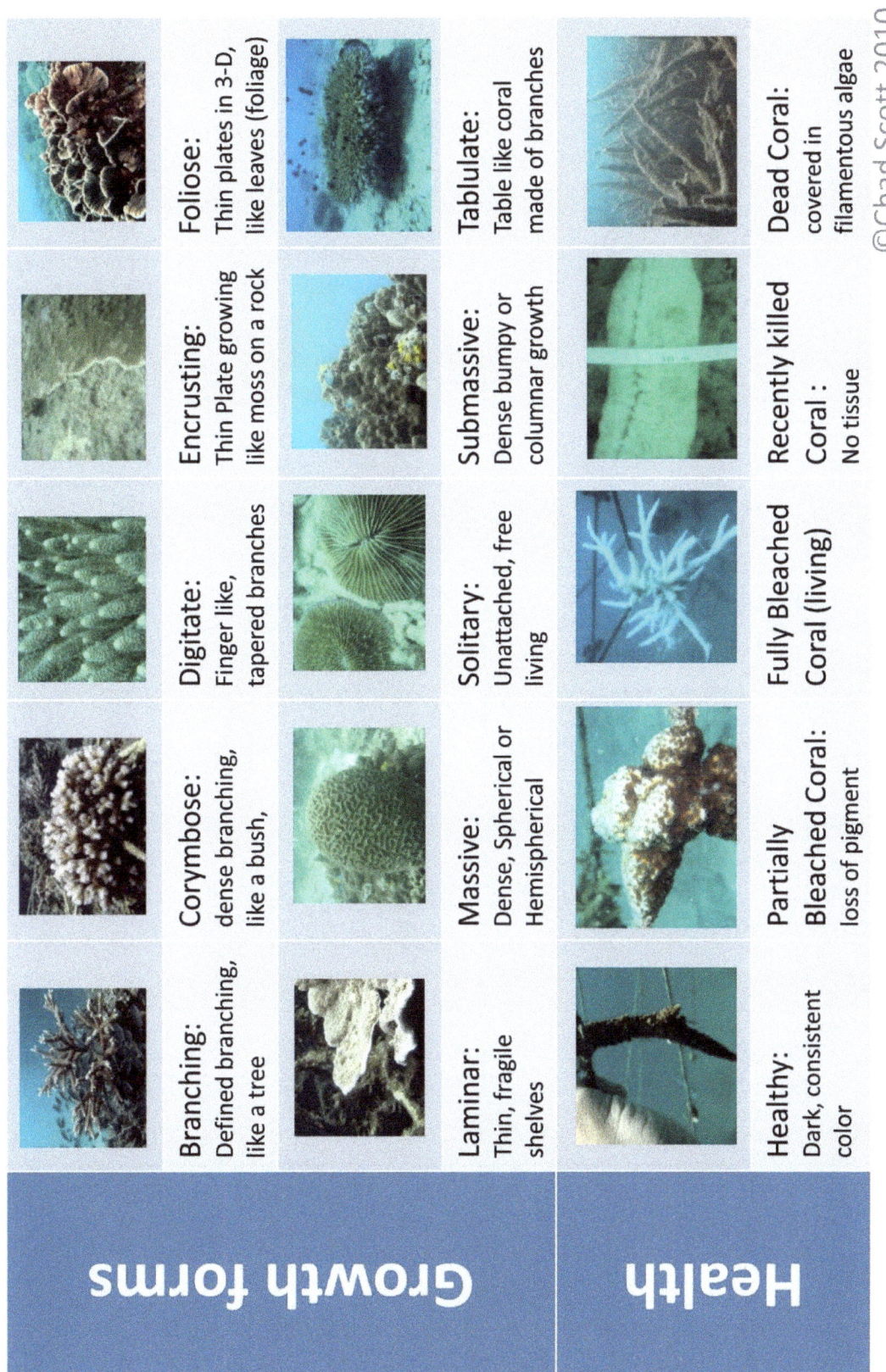

© Chad Scott 2010

Ecological Monitoring Program Manual

Appendix D
List of Coral Genera (for Advanced EMP Surveyors)

Genus	Code
Acropora	ACRO
Acanthastrea	ACAN
Alveopora	ALVEO
Astreopora	ASTREO
Caulastrea	CAUL
Coscinarea	COSI
Ctenactis	CTEN
Cyphastrea	CYPH
Danafunia	DANA
Diploastrea	DIPLO
Echinophyllia	ECHINPH
Echinopora	ECHINPO
Euphyllia	EUPHY
Favia	FAV
Favites	FAVIT
Fungia	FUNG
Galaxea	GALA
Goniastrea	GONIA
Goniopora	GONIO
Hydnophora	HYDNO
Herpolitha	HERP
Leptoria	LEPTO
Leptoseris	LEPTS
Lobophyllia	LOBO
Merulina	MERU
Millipora	MILL
Montastrea	MONTA
Montipora	MONTI
Oulophyllia	OULO
Pachyseris	PACHY
Pavona	PAV
Pectinia	PECT
Platygyra	PLATY
Pocillopora	POCI

Genus	Code
Podabacia	PODA
Porites	POR
Psammocora	PSAM
Sandalolitha	SAND
Seriatopora	SERIA
Siderastrea	SIDER
Stylocoeniella	STYLO
Tubeastrea	TUBEA
Turbinaria	TURB
Unknown	UKN

Appendix E
Example Compromised Coral Health Survey Form

Name:
Date: Depth: Temp:
Location: Conditions:

Family	Genus	Shape	Healthy	SP	TA	MA	CB	COT	DRUP	DAMS	TRIG	PRT	BTFY	BBD	YBD	BrB	SEB	WS	GA	PGA	TMD	SDS	PHY	UBP	OTH
										PREDATION					**DISEASE**						**OTHER**				
ACRO	Acropora	B																								
		C																								
		D																								
		T																								
	Montipora	M																								
		E																								
AGAR	Astreopora																									
	Leptoseris																									
	Pavona																									
DEND	Turbinaria																									
DIPLO	Diploastrea																									
FAVI	Favia																									
LOBO	Lobophyllia																									
MERU	Oxypora																									
	Cyphastrea																									
	Echinopora																									
	Favites																									
	Leptoria																									
	Merulina																									
	Pectinia																									
	Platygyra																									
POCI	Pocillopora																									
POR	Porites	M																								
		B/S																								
OTH																										

APPENDIX F
CORAL BLEAACHING EXAMPLE SURVEY SLATE

Name
Date
Location
Temp
Conditions

Family	Genus	Shape	Healthy	PBL			FBL	Mortality		Coral Watch			Notes
				%P	%M	%W	%RKC	%DC	LI Code	LI #	DI Code	DI #	
ACRO	Acropora	B											
		C											
		D											
		T											
	Montipora	M											
	Astreopora	E											
AGAR	Leptoseris												
	Pavona												
DEND	Turbinaria												
DIPLO	Diploastrea												
FAVI	Favia												
LOBO	Lobophyllia												
MERU	Oxypora												
	Cyphastrea												
	Echinopora												
	Favites												
	Leptoria												
	Merulina												
	Pectinia												
	Platygyra												
POCI	Pocillopora												
POR	Porites	M											
		B/S											
OTH													

Appendix G
Dangerous Marine Animals of the Indo-Pacific

Here you can find a list of the animals that you should be careful around. This is provided not to scare you, but to improve your safety by informing you what to watch out for, and what to do if you encounter some of these species. This is not a complete list, your instructor may tell you about other threats in your particular area. Please note that we have never had a conservation student injured by any of these marine species, and want to keep it that way. So, learn what to avoid and stay safe while you have fun enjoying your course!

General notes on marine stings, bites, and toxins

Most dangerous animals encountered while diving do not 'attack' humans, and are easily avoided by practicing good buoyancy and not contacting the reef. In general, a hands-off policy will keep you safe at all times. However, in Conservation Diving we must sometimes handle marine life or kneel on the sandy sea beds while doing artificial reef work, so divers must always pay attention. The following rules will ensure your safety:

- Know which animals are dangerous and how to avoid them
- Know what to do if stung/bit
- Never touch anything unless instructed to do so
- If you don't know what it is, don't touch it
- Never collect marine life or shells

Also, marine bites or stings while diving carry the added risk of panic and diving related injuries, so if you are stung or bitten follow these steps:

1. **Don't Panic** – Stop what you are doing, breath, and think before doing anything else. DO NOT ASCEND
2. **Signal to your buddy/dive leader** that you are having a problem
3. If possible, also **point out the animal to your dive buddy/leader** so they know what precautions and treatment should be taken
4. **Perform a controlled ascent**, with your buddy, of no more than 9 meters per minute (DO NOT do a safety stop)
5. Inflate your BCD, signal to the boat, and request assistance

As with any problems you could potentially have while diving, the most important thing is to always stay calm. Little problems can quickly become big problems when people panic. To prevent panic, know yourself, know your limits, and have a plan for what to do in case of injuries and emergencies.

Stop – Think – Plan - Act

Jellyfish

There are several species of jellyfish which are a danger, most are in fact not harmful. However, in recent years there have been increased sightings of several poisonous species of jellyfish, including Box Jelly fish (*chirodropids* and *carybdeids*). The Portuguese Man-of-War is also a danger in many parts of the world's oceans.

Frequency of Encounters:

Rare in most areas, although speak with locals and your instructor to find out their seasonality and if it is a particularly dangerous time of year.

What to Watch out for:

Jellyfish drift slowly through the water, and are thus easily avoided by divers paying attention. In most cases, stings result from people wading or swimming without a mask, or trying to pick up or handle a jellyfish, thinking it is not dangerous. If you encounter a jellyfish under water, maintain a safe distance. Also, the use of rash guards, stinger suits, or wetsuits is strongly encouraged while diving, snorkeling, or swimming.

What to do if Stung:

Do not remove tentacles with your bare hands, instead flush them off with sea water or use tweezers. Immediately wash the affected area with vinegar and seek medical attention. Oxygen should also be administered and heart rate should be constantly monitored.

Cone snails

Cone snails are a very diverse group of marine gastropods. Although most are not poisonous, the ones which are contain some of the most dangerous and potent toxins on the planet. All species should thus be avoided and never handled.

Frequency of Encounters:

Common to many sites throughout the Indo-Pacific, especially deeper than 12 meters, occasionally also wash up on the beach.

What to Watch out for:

The cone snail has a 'harpoon' which it uses to kill and eat fish or crustaceans, the barb of this 'harpoon' contains the powerful neurotoxins. The cone snails will sting you if you handle them, so never pick-up or touch a cone snail, even if you think it is dead.

What to do if Stung:

There is no anti-venom for cone snail toxins, follow the steps at the beginning of this chapter and seek medical attention immediately. Hot water can be applied to decrease the potency of the toxin, but should not interfere with or delay access to professional medical care.

TOXIC SEA URCHINS

The collector urchin, also known as the flower urchin (*Toxopneustes pileolus*).

Frequency of Encounters:

Common in reef/rubble areas throughout the Indo-Pacific

What to Watch out for:

The collector urchin is hard to spot as it often covers itself in rubble, sponges, and other debris. Also have a good look before kneeling down in the sand, or around the giant clam nurseries, and while collecting coral fragments for the coral nurseries.

It is, in fact, not the spines which are dangerous, but the round suction cups, called pedicellariae, they are actually little claws which close up, injecting venom into whatever has tried to touch it. In humans, effects begin several seconds after being injected, and generally include excruciating pain and paralysis, lasting up to 6 hours. Infections and secondary symtptoms can last much longer.

What to do if Stung:

Notify your buddy and immediately begin a controlled ascent to the surface. Inflate you BCD and call for help. Swimmers are thought to have drowned after being paralyzed by the toxin, so be sure somebody knows right away that you require assistance.

Stonefish/Scorpionfish

There are several species of stonefish and scorpionfish, all of which are thought to be poisonous. The most common species encountered are the raggy scorpionfish (*Scorpaenopsis venosa*) and the Devil Stinger or Indian Walkman (*Inimicus didactylus*). Stonefish are camouflaged to closely resemble features on the reef, and are difficult to spot.

Frequency of Encounters:

Common, seen every few dives (if you have a good eye for them).

What to Watch out for:

The fins on the back of the fish contains the toxin, which is injected when a person steps onto or grabs the fish. As one of the most venomous fish in the world, the toxin causes long-lasting pain and secondary problems involved with swelling and infections. In some cases the sting can also be fatal. Most stings occur from people walking on the reef, but occasionally divers kneel on or put their hands onto the fish. Avoid this by refraining from walking on the sea bed and not grabbing onto rocks or dead coral while diving. During conservation activities always check coral fragments and structures before you touch them, and keep an eye out for these fish. If you do happen to see one on an artificial reef, be sure to let the other divers working with you where it is. The sign for stonefish is to point with your little finger or a closed fist (danger).

What to do if Stung:

There are no anti-venoms for stonefish, although medical facilities can treat some of the symptoms. Victims have reported unbearable levels of pain associated with stings, which can cause fainting or shock. In some cases symptoms can last several months. It is vital that you remain calm and still perform a controlled ascent from the dive, signal or get quickly to the boat where you can be treated for shock and transported to a hospital.

CROWN OF THORNS STARFISH

Acanthaster spp.

Frequency of Encounters:

Very common (multiple individuals seen on each dive)

What to Watch out for:

The spines covering the Crown of Thorns Starfish are not technically poisonous, but do cause great pain and discomfort if they enter the body. Some victims may also experience anaphylactic shock if they are allergic to the sting. Avoid handling these starfish, and watch your buoyancy and you will have nothing to worry about.

What to do if Stung:

Keep calm, although the sting is very painful, it is generally not life threatening. Massage the affected area, starting from close to the heart and pushing towards the wound to force out any toxins or pieces of the spines. Perform a controlled ascent and the put the affected area in hot water on the boat for 20-30 minutes.

TRIGGER FISH

Multiple species, the most problematic of which is the Titan Triggerfish (*Balistoides viridescens*)

Frequency of Encounters:

Common (seen on most dives, but rarely a problem)

What to Watch out for:

The titan triggerfish is a large fish that will become quite territorial and defensive during nesting season. Their territory can be imagined as an cone, which starts from the nest and gets wider up towards the water's surface. Generally, the triggerfish will begin swimming erratically and putting up its 'trigger' to let divers know they are too close. Divers who fail to recognize these signs, or are not looking out for the fish, can be attacked. Generally the fish will ram or hit the divers, but in some cases can also bite. This is not life-threatening, but can be quite scary.

What to do if attacked:

First, do not panic or go up. Fast movements look threatening to the fish, and going up not only puts you in danger of decompression injuries, but also puts you more into their territory. Turn and face the fish, put your fins between you and the fish, and begin swimming backwards to get out of their territory.

Ecological Monitoring Program Manual

Glossary

Abiotic
Non-living, not derived from living organisms (see also *Biotic*).

Abundance
The quantity or amount of something (ie. the number of animals in a population).

Anthropogenic
Originating from human activities (ie. environmental pollution).

Asexual Reproduction
Reproduction without a partner. Involves cloning or budding, to form offspring with identical DNA to the parent (see also *Sexual Reproduction*).

Axial polyp
Coral Polyp orientated to the direction of growth. The polyp at the branch tip (see Acropora Corals for example.)

Benthic
Living on in the substrate/sea bottom. Often describes non-swimming animals (sea stars, urchins, worms, etc).

Biodiversity
The variety of life in an ecosystem, the number of species or phyla in a region, etc.

Biology
The study of living organisms

Bio-mineralization
The process by which living organisms produce minerals, often to harden tissues or create structure.

Biotic
Living, relating to or resulting from living things.
(See also *Abiotic*).

Bivalves
An aquatic mollusk that has a body enclosed in two hinged shells (i.e. clams, mussels oysters, etc.)

Calice
Cup-shaped depression on the corallite surface.

Cambrian Explosion
The rapid diversification of multi-cellular animal life resulting in the appearance of almost all modern animal phyla (Beginning around 530 million years ago).

Carnivores
An animal that feeds on flesh (predators).

Cartilaginous Fish
A fish whose skeleton consists primarily of cartilage (ie. Sharks, rays, skates, etc).

Cephalopods
An aquatic mollusk that has feet originating near or at the head (ie. Octopus, squid, cuttlefish, etc).

Coenosarc
The soft tissue that links colonial coral polyps together.

Commensalism
An association between two organisms in which one benefits and the other derives neither benefit nor harm.

Corallite
Skeletal structure of an individual polyp (either solitary or within a colony).

Corallivores
An animal which eats coral polyps (COTS, Drupella, Butterfly fish, etc).

Crustaceans
Class of arthropods including lobster, crabs, shrimps, Krill, decapods, etc.

Cryptic
Being hidden, indescript, or difficult to locate (i.e. Drupella snails which hide under corals during the day-time.)

Cyanobacteria
A type of microorganism that are related to bacteria but are capable of photosynthesis.

Demersal
A vertebrate, mostly fish, which live close to the sea bed (ie. Groupers)

Dendritic
Growing in a tree shaped form

Dinoflagellate algae
A single celled organisms with two flagella (looking like tails) (see also Zooxanthallae).

Dynamic substrate
A substrate or sea bottom which is unsecured or prone to movement (ie. Silt, sand, rubble).

Ecosystems
A biological community of interacting organisms and their physical environment.

Electroreception
The detection of electrical fields or currents by aquatic animals.

Eukaryotic cell
A biological cell which contains complex internal structures enclosed within membranes (including a nucleus). The cells which make up the bodies of all plants and invertebrates (see also *Prokaryotic cell*).

Fecundity
The ability to reproduce, fertility.

Filamentous algae
Single celled algae cells which form long chains, threads, or filaments.

Fusiform
Elongated and tapering at each end (like barracuda).

Gastropods
A class of mollusks, typically having a one piece shell and a stomach orientated foot.

Growth form
The overall shape of a coral colony (ie branching, table, mushroom, etc.)

Herbivores
An animal which eats pants and vegetation or algae (primary consumer).

Hermatypic corals
Reef building corals which are characterized by the presence of symbiotic algae within their tissues.

Holobiont
The collective community of a coral host and its microbial symbionts.

Indicator species
A species which, due to its biological or ecological functions, can provide researches with information on ecosystem health and threats.

Inorganic
Not derived from natural growth (ie. minerals, rocks. etc.)

Invertebrates
An animal lacking a backbone.

Macro-Algae
Macroscopic, multicellular algae (ie seaweed and kelp).

Metadata
A set of data which describes and gives information about other data.

Micro-invertebrates
Microscopic (not visible to the naked eye) invertebrates (ie. zooplankton)

Multicellular
An organism or plant which is comprised of many cells.

Mutualism
A symbiosis which is beneficial to both organisms involved.

Nematocysts
A specialized cell in the tentacles of members of the Cnidarians phylum containing a barbed or venomous coiled thread.

Notochord
A cartilaginous skeletal rod supporting the body in all embryonic and some adult chordates.

Octocorals
A class of corals containing tentacles in multiplications of 8, generally are soft bodied or non-reef building).

Oligotrophic
Water which is relatively low in nutrients and has abundant levels of oxygen

Organic
Of, relating to, or derived from living matter.

Parasitism
A symbiosis between two organisms in which the symbionts receives benefits while the host is at a loss.

Pelagic
Of or relating to the open sea (ie deep sea fish, tuna, etc.)

Perciform
One of the largest natural groups of fishes which have a perch or tuna like appearance.

Photosynthesis
The process by plants and other organisms in which they use sunlight to synthesize energy from carbon dioxide and water.

Phylogenetic tree
A branching diagram showing inferred evolutionary relationships between animal phyla.

Phylum
A taxonomic category that ranks below Kingdom (Plants, Animals, Fungi) but above Class.

Plankton
Organisms drifting or floating in the water, those which lack ability to swim against currents. Often microscopic, but that is not a defining characteristic (large jelly fish can be planktonic).

Porous
Having minute holes through which water or air can pass.

Positive Feedback Loops
A process by which an effect will act to amply the magnitude of future effects.

Prokaryotic cell
A biological cell which lacks an enclosed nucleus, tends to be on the order of 10,000-100,000 times smaller than an Eukaryotic cell (ie. bacteria, archaebacteria.)

Protienacious skeleton
A skeleton consisting or derived from protein molecules (ie sea fans and sea whips).

Reef Resilience
The ability of a reef ecosystem to absorb, resist, or recover from a disturbance. Also the ability to adapt to changes with a relatively low loss to abundance or biodiversity.

Salinity
The amount of dissolved solids in sea water (generally around 32-35 PPM, or about 3.1-3.8%)

Scleractinian
The order of Anthazoa containing the hard or stony corals.

Secondary Consumer
Predators, or animals which feed on herbivores.

Serpentine
Of or relating to a snake shape.

Sessile
Non-moving, or attached organism (ie. barnacles or tunicates).

Stromatolites
Layered accretionary structures produced by the bio-mineralization activities of unicellular organisms (primarily cyanobacteria).

Substrate
The bottom layer, or surface material on or from which an organisms lives, grows, or receives nourishment.

Symbiosis
The interaction of two different organisms living in close physical association.

Top-Predator
The highest predator in a trophic structure. An organism which has not other predators above it.

Tridacna Family
Family of marine bivalves containing the giant and boring clams.

Trophic Level
Each of the several hierarchal levels in an ecosystem based on energy flow through production and consumption.

Tunicates and Ascidians
Small marine invertebrates which have a rubbery or hard tissue and an in-current and out-current siphon to filter feed. (ie. sea squirts).

Unicellular
Organisms which are comprised of only a single cell (see also Multicellular).

Vertebrates
Class of Chordate animals which are characterized by having a backbone and a post natal tail either in the embryonic and sometimes adult form.

Xenic nutrients
Excess or anthropogenic derived nutrients in a living system.

Zooplankton
Plankton which belongs to the animal kingdom (non-photosynthetic).

Zooxanthallae
Flagellated single celled organisms that are endosymbionts of various marine animals.

LIST OF FIGURES AND PHOTO CREDIT

All photos and graphics were the work of the author, unless otherwise noted below:

- Chapter 1
    - Coral close-up: Heike Schwermer
    - Coral Anatomy drawing: Adapted from Christine Elder, 2006
- Chapter 3
    - Nudibranch, Cuttlefish: Pau Urgell
    - Cushion Star: Rahul Mehrotra
    - Black, Marbled, and Pink Fish Sea Cucumbers: Nathan Cook
- Chapter 4
    - Shark 1: Sirachai Arunrugstichai
    - Shark 2, Ribbon Ray, bumphead parrot, Snapper, Surgeonfish, Clingfish, Giant Moray, Seasnake: Pau Urgell
    - Clingfish: Spencer Arnold
- Chapter 9
    - Goatfish: Kirsty Magson

REFERENCES AND WORKS CITED

Brown, B.E. 1997. Coral Bleaching: Causes and Consequences. *Coral Reefs* 16, Suppl.:S129-S138.

Bruno, J.F., Selig E.R. 2007. Regional Decline of Coral Cover in the Indo Pacific: Timing, Extent, and Subregional Comparisons. *PLoS ONE* 2(8).

Costanza, R. 1997. The Value of the World's Ecosystem Services and Nature Capital. *Nature London*. Vol. 387, No. 15, pp. 253.

Castro, P., Huber, M. 2007. **Marine Biology: 6th Edition**. McGraw Hill. New York, NY.

Chavanich, S., Viyakarn, V., Adams, P., Klammer J., Cook, N. 2012. Reef Communities After the 2010 Mass Coral Bleaching at Racha Yai Island in the Andaman Sea and Koh Tao in the Gulf of Thailand. *Phuket Mar. Biol. Cent. Res. Bull.* 71: 103-110.

Garces, L.R. 1992. Coral reef management in Thailand. Coral reef management in Thailand. Naga, the ICLARM Quarterly, 15(3), pp. 40-42.

Hein M, Lamb JB, Scott CM, Willis BL (2014) Assessing the potential for marine protected areas to ameliorate coral health in one of the world's scuba diving hotspots. *Marine Environmental Research*.

Hoegh-Guldberg, O. 1999. Climate Change, Coral Bleaching and the Future of the World's Coral Reefs. *Marine and Freshwater Research*, CSIRO Publishing, Vol. 50, pp. 839-66.

Hoeskema, B.W., Scott, C.M., True, J.D. 2013. Dietary Shift in Coralivorous *Drupella* snails following a major bleaching event at Koh Tao, Gulf of Thailand. *Coral Reefs* Vol 32, Issue 2, pp. 423-428.

Knowlton N., Jackson J.B.C. 2008. Shifting Baselines, Local Impacts, and Global Change on Coral Reefs. *PLoS Biol* 6(2): e54. doi:10.1371/journal.pbio.0060054

Knowlton, N. 2001. The Future of Coral Reefs. Presentation at the National Academy of Sciences. *PNAS* Vol. 98, 10:5419-25.

Koop K., Booth D., Broadbent A., Brodie J., et. al. 2001 ENCORE: the effect of nutrient enrichment on coral reefs – Synthesis of results and conclusions. *Marine Pollution Bulletin* 42: 91-120

Lamb JB, True JD, Piromvaragorn S, Willis BL (2014) Scuba diving damage and intensity of tourist activities increases coral disease prevalence. *Biological Conservation* 178:89-96.

Larpnun, R., Scott, C.M., Surasawadi, P. 2011. Practical Coral Reef Management on a small island: Controlling Sediment on Koh Tao, Thailand. Pp94-95 in Wilkinson C., Brodie J. (eds.) **Catchment Management and Coral Reef Conservation**. *Global Coral Monitoring Network and Reef and Rainforest Research Centre*. Townsville, Austrailia, 120p.

Ministry of the Environment, Japan. 2010. Status of Coral Reefs in East Asian Sea Region: 2010. *Global Coral Reef Monitoring Network*, 121pp.

Nichols, R. 2013. Effectiveness of artificial reefs as alternative dive sites to reduce diving pressure on natural reefs, a case study of Koh Tao, Thailand. Bsc. Thesis in Conservation Biology, University of Cumbria, Cumbria, UK. Print.

NOAA. 2011. NOAA: 2010 Tied for warmest year on record. 12 January 2011. Obtained from http://www.noaanews.noaa.gov/stories2011/20110112_globalstats.html on 24 February 2013.

Pandolfi, J.M, Bradbury, R.H., Sala, E., Hughes, T.P., *et al.* 2003. Global Trajectories of the long-term Decline of Coral Reef Ecosystems. *Science Magazine,* Vol. 301, pp. 955-958.

Phongsuwan N., Chankong, A., et al. 2013. Status and changing Patterns on Coral Reefs in Thailand during the last two decades. PMBC, 18pp

Phillips, W.N., Scott, C.M., Zahir, D. 2010. Community Monitoring and Assessment of the Reefs of Koh Tao, Thailand. Proceeding of the the 2nd Asia Pacific Coral Reef Symposium. Marine Biodiversity Research Group, June 20-24, 2010. Phuket, Thailand.

Platong, S., Chaloem, S., Charoenmart K., *et al.* 2012. **Strategic Plan: Integrated Coastal Management for Koh Tao**. *Center for Biodiveristy of Peninsular Thailand, Prince of Songkla University*, Hat Yai, Thailand.

PMBC. 2010. Coral Bleaching in Thailand in 2010. Unpublished report. Phuket Marine Biological Center.

Raymundo, L.J., Couch, C.S., Brukner, A.W., Harvell, C.D., *et al.* 2008.**Coral Disease Handbook: Guidelines for Assessment, Monitoring & Management.** *Coral Reef Targeted Research & Capacity Building for Management Program.* Townsville, Australia.

Satapoomin, U. 2000. A Preliminary Checklist of Coral Reef Fishes of the Gulf of Thailand, South China Sea. Phuket Mar Biol. Cet. res. bull. 48(1).

Scaps, P. 2006. Eight New Records of Coral Reef Fishes from the Gulf of Thailand, South China Sea. Phuket Mar Biol. Cet. res. bull. 67: 53-62.

Scaps, P., & Scott, C. (2014). An update to the list of coral reef fishes from Koh Tao, Gulf of Thailand. Check List, 10(5), 1123-1133.

Scott, C.M. 2008. Loss of Tanote Bay: Analysis and Request for Assistance. Save Koh Tao, 6pp.

Scott, C., Phillips, W.N. 2010. A Sustainable Model for Resource Management and Protection Achievable Through Empowering Local Communities and Businesses. Proceedings of Ramkhamhaeng University International Research Conference 2010, January 13-14, 2011, Bangkok, Thailand.

Walther, G-R., Post, E., Convey, P., Menzel, A., et. al. 2002. Ecological Responses to Recent Climate Change. . *Nature.* 416:389-395.

West, J.M., Salm, R.V. 2003. Resistance and Resilience to Coral Bleaching: Implications for Coral Reef Conservation and Management. *Conservation Biology.* Vol 17. 4:956-67

Wetterings, R. 2011. A GIS-Based Assessment to the Threats to the Natural Environment on Koh Tao, Thailand. *Kasetsart J. (Nat. Sci.)* 45 743:755

Wilkinson, C.R. 1999. Global and Local Threats to Coral Reef Functioning and Existence: Review and Predictions. *Marine and Freshwater Research,* CISIRO Publishing, Australia. 50:867-78.

Wilkinson, C.R. 2008. Status of Coral Reefs in the World: 2008. *Global Coral Reef Monitoring Network and Reef and Rainforest Research Centre,* Australia.

Wilkinson, C.R. 2010. Coral Reefs of the Asia-Pacific Region: Status and Trends and Predictions for the Future. Abstract for KeynoteAddress Proceeding of the the 2nd Asia Pacific Coral Reef Symposium. Marine Biodiversity Research Group, June 20-24, 2010. Phuket, Thailand.

YeeminT., Mantachitra V., Platong S., Nuclear P., Klingthong W., Suttacheep M. 2012. Impacts of Coral Bleaching, Recovery, and Management in Thailand. Proceedings of the 12 International Coral Reef Symposium, Cairns, Australia, 9-13 July 2012.

BE SURE TO CHECK OUT OUR OTHER EXCITING MARINE CONSERVATION COURSES AND CERTIFICATIONS, INCLUDING:

- Advanced Ecological Monitoring Program
- Coral Nursery Theories and Techniques
- Artificial reef Theories and Techniques
- Sea Turtle Ecology and Head-Starting
- Seahorse Ecology and Monitoring
- Giant Clam Nurseries and Population Studies
- Coral Spawning and Larval Culturing Program
- Shark Ecology and Populations Studies
- Nudibranch Ecology and ID
- Mooring Line Maintenance and Installation
- Coral Taxonomy and Identification
- Mineral Accretion Device Basics and Techniques
- Coral Predators: Population Monitoring and Management
- And More to Come...

Don't leave just yet. . . .

We want to hear from you. Please let us know how you liked this manual by emailing us at **Info@ConservationDiver.com**. Please let us know what was helpful about this manual, and what can be improved. We put out a new edition every few years and would love to know how to make this manual more helpful or applicable to your particular area.

If you have interesting observations or findings during your EMPs, you can also email us that info and we will do our best to record it or explain it.

Also, just a reminder, please help to support these great programs and get others involved in citizen science activities:

iSeahorse.org

iSeahorse.org is an organization dedicaed to increasing awareness and collecting data on seahorses around the globe. If you are lucky enough to see a seahorse while diving, snorkeling, or on the beach please take a picture and report your sighting to them. Its free to sign-up, and anybody can participate.

Coralwath.org

CoralWatch.org is a collaboration of several universities to create an easy, yet valuable, system to report coral health and bleaching. By signing up on their website you can receive a packet of information and an underwater survey slate to get you started. After that you can collect and report data on coral health for every dive you do, even fun dives.

whaleshark.org

Whaleshark.org is an organization that uses photographs to identify and track whalesharks around the world. You can learn from their website how to take the photos, and be prepared for that big moment. This project is important to understanding the populations and migrations of the biggest fish in the sea.

Lastly, if you love our programs and want to do more to support them, we have a donation link on our webpage.

CONSERVATION DIVER

Continue your Conservation Diving Adventure

Check out some of the great learning materials and other gear available on our website

www.ConservationDiver.com

Fish, Invertebrate, & Substrate Underwater ID Slates
+ Bonus Turtle & Seahorse Cards

Hand Signal Charts

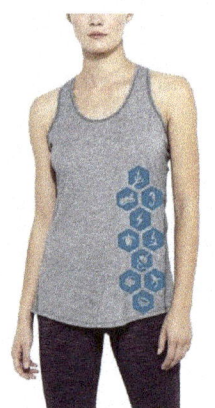

Women's Shirts

Thin Hoodies

Hats

And a whole lot more

www.ingramcontent.com/pod-product-compliance
Lightning Source LLC
Chambersburg PA
CBHW041657040426
R18086800001B/R180868PG42333CBX00010B/11